智能交互服装的融合设计与人文评价

Toward the Integration of Cyber-Physical Systems with Emotional Evaluation: A Perspective on Interactive Clothing Design

王伟珍　著

中国纺织出版社有限公司

内　容　提　要

数智工程与时尚产业的深度融合，再次迭代了时尚服装领域的设计思维与产品开发方法。随着智能技术与可穿戴概念的演进，以智能交互服装为载体的人文设计正转向"人—智能体—社会"交互协作的设计新范式。本书从智能服装的源起、概念、原理、技术、案例等层面系统阐述作者的逻辑观点和实践经验等，读者可以从中探赜设计工学的深度思考，同时也可以深入了解可穿戴交互时尚产业的发展趋势，并获得在智能交互时尚设计方面的启示。

本书适合纺织服装、可穿戴时尚设计、交互设计等交叉学科专业的师生阅读，以及对趋势发展感兴趣的读者收藏。

图书在版编目（CIP）数据

智能交互服装的融合设计与人文评价：英文 / 王伟珍著． -- 北京：中国纺织出版社有限公司，2024.9.
ISBN 978-7-5229-2223-2

Ⅰ．TS941.2-39

中国国家版本馆 CIP 数据核字第 20248KV257 号

责任编辑：张艺伟　魏　萌　　责任校对：高　涵
责任印制：王艳丽

中国纺织出版社有限公司出版发行
地址：北京市朝阳区百子湾东里A407号楼　邮政编码：100124
销售电话：010—67004422　传真：010—87155801
http://www.c-textilep.com
中国纺织出版社天猫旗舰店
官方微博 http://weibo.com/2119887771
天津千鹤文化传播有限公司印刷　各地新华书店经销
2024年9月第1版第1次印刷
开本：787×1092　1/16　印张：8.25
字数：220千字　定价：78.00元

凡购本书，如有缺页、倒页、脱页，由本社图书营销中心调换

Preface

  Revolutionary changes have been occurring at an unprecedented rate in the fields of clothing. The interactive clothing, a topic on which there is little previous research, an evolutionary branch of smart clothing in the field of information science, which emphasizes the function of social symbols that mutual interaction or communication between the wearers and their environment based on the integration of information science and traditional clothing. Combining Cyber-Physical Systems (CPS) with clothing engineering designing to input a certain physical signal into clothing, the interactive clothing can output a certain social symbol that people or clothing environment can perceive and generate corresponding interaction. This kind of social symbol is the expression form of the interactive clothing as the interactive medium for people to interact with the environment. With the gradual integration of the Internet of Things (IoT) and CPS into people's daily life and the growing development of smart textile technology, interactive clothing will play an increasingly important role in future interpersonal communication and interaction.

  Despite the impressive track record in smart clothing research and development, we still have to confront with several current research dilemmas or development bottlenecks from correlative humanities science and information technology respectively. Researchers have generally focused on high-tech approaches to implement smart clothing design with multifunction. Nevertheless, the complex sociological attributes of clothing, i. e., its interactive symbolism and properties of emotional expression, design hierarchy and design aesthetics, and other aspects should not be ignored.

  The overall purposes of this book are to bridge the gap between CPS applications in the field of information science and emotional evaluation in the field of humanistic science, to minimize the unbalance between humanistic emotion and wearable smart technologies for interactive clothing innovation, to investigate how the transformation could be realized from information to knowledge in the process of interactive clothing design, to establish a basic framework of design principle and design evaluation criteria system for interactive clothing, and to suggest practical implication for interactive clothing design.

  To attain the above objectives, this book will answer one major research question: What happened to the interactive clothing when the social semantic transformed from information stage to knowledge stage in the architecture of CPS? In addition, three subsidiary research questions (SRQ), according to soft system methodology (SSM), i. e., the "WHY", "HOW" and "WHAT" as follows. SRQ 1: Why does the Research and Development (R&D) of interactive clothing worn in daily life need to integrate smart technology with emotional design and humanistic evaluation simultaneously? SRQ 2: How could the technology transformation be realized among the signals in the field of information physics and the symbols in the various dimensions of humanity and society? SRQ 3: What kinds of the design principles of interactive clothing should be valued in the context of CPS's growing prosperity?

The author's practical research on this topic began in 2012, and progressive results were gotten during three stages. The first stage was in 2013, two prototypes of infant's smart clothing were developed. As a result, two Chinese Utility Model Patents have been granted, which illustrates the feasibility and rationality of this study's entry point. The second stage was the study of interactive couple clothing prototype, which is a kind of clothing media that can transfer some interpersonal relationships, completed in 2016-2017. Further, the diversified interaction results have been achieved through a three-piece set of interactive parent-child clothing prototype in the third stage from 2017.

In theoretical research, by literature review and literature integration, the psychology, sociology and design hierarchy of clothing, as well as Data-to-Information-to-Knowledge-to-Wisdom (DIKW) model and CPS architecture are deduced and summarized from the perspective of humanities and technology respectively to analyze the humanistic and technical attributes of interactive clothing. The Cyber-Physical-Clothing Systems (CPCS) model of the technical development process of interactive clothing is created, and the validity of the CPCS model is verified by the method of prototype development and Kansei Engineering evaluation. Through Kansei Engineering evaluation, the design elements and evaluation criteria of interactive clothing are also extracted. The results found that the research on interactive clothing should integrate the two opposing perspectives of humanities and technology, and bridge their gaps from an interdisciplinary perspective in the process of prototyping and evaluating.

The main differences among the contents of this book and previous studies by other scholars are as follows: ①New prospective, i.e., this book pioneered a new field of smart clothing research, namely interactive clothing; ②New approach, i.e., this book introduced CPS technology into clothing development from the approach of art design; ③New system, i.e., a research framework for interactive clothing was initially established.

The main original contributions of this book are as follows: ①The concept of interactive clothing was a pioneer definition; ②From the perspective of humanities and technology respectively, the models of studying the attributes of interactive clothing are built to guide the research path; ③Integration of CPS, DIKW and other information technology and theories to create a CPCS architecture model to guide the prototype development of interactive clothing; ④The 18 "C" Design Principles of interactive clothing are revealed; ⑤The criteria framework of interactive clothing design evaluation is created to encourage the object of evaluation to develop in the right direction and objectives. Therefore, the knowledge framework associated with interactive clothing is preliminarily established in the category of Knowledge Science.

This book is a phased achievement of the Social Science Planning Fund of the Ministry of Education of China's project (21YJAZH088), the key research project of the Education Department of Liaoning Province, China (LJKZZ20220069).

Wang Weizhen
Dalian Polytechnic University
May 2024

引言

聚智新科技范式，赋能新设计产业。数字赋能时代，服装技术领域也以前所未有的速度更新迭代。数智工程与时尚产业的深度融合，再次迭代了时尚服装领域设计思维与产品开发的方法。智能技术与可穿戴概念的演进，正全面冲击传统服装设计的基础、对象、方法和流程，以智能交互服装为载体的人文设计正转向"人—智能体—社会"交互协作的设计新范式。

智能交互服装是智能服装在信息科学领域的一个进化分支，它强调基于信息科学技术（网络物理系统，CPS）与传统服装相融合的社会符号功能，即穿着者与环境之间的彼此感知互动或交流。这种社会符号是交互服装作为着装者与环境互动的媒介表征。随着CPS逐渐融入日常生活以及智能纺织技术的发展，交互服装将在未来的人际互动中发挥愈加重要的作用。

本书将回答一个主要问题：在CPS架构中，交互服装的社会语义从信息阶段向知识阶段的转变过程中发生了什么？此外，根据系统方法论将回答三个次要问题：为什么针对日常生活中穿戴的交互服装研发，需要同时将智能技术与情感设计、人文评价整合？技术方面，如何实现从信息物理领域的信号向人类社会多维度的符号转化？在CPS日益繁荣的背景下，交互服装的设计原则有哪些？

本书主要内容包括：①智能交互服装的概念；②从人文与技术的跨领域角度出发，构建研究交互服装属性的模型；③整合CPS、DIKW等信息技术和理论，打造"人—服装—环境的网络系统"架构模型，引导交互服装的原型开发；④提炼从物理信号输入到社会符号输出的流程，指导交互服装社会语义交互的多样化设计；⑤揭示交互服装的设计原则；⑥创建交互服装设计评价的标准框架。

本书立足于新兴交叉学科领域，秉承设计工学（Design Engineering）理念，打破服装设计方法受传统设计学困囿的艺术思维范式，基于科学、技术、工程、艺术、数学（STEAM）相融合的跨学科视角，通过服装设计学、信息物理系统（CPS）、感性工学（Kansei Engineering）三个理论层面，围绕智能交互服装开展理论论述和实践案例分析，向读者系统介绍"人—服装—环境系统"背景下智能交互服装的理论概念与研究方法。

本书是教育部人文社会科学研究规划基金项目《人工智能服装设计的思维范式与生成机理研究》（21YJAZH088）、辽宁省教育厅重点攻关项目《基于图像翻译和GAN神经网络的服装智能设计机理与方法研究》（LJKZZ20220069）的阶段性成果，也是作者开展人工智能设计研究的基础成果。

王伟珍
大连工业大学服装人因与智能设计研究中心
2024年5月

Contents

Chapter 1
Introduction ··· 1
  1.1 Research background ·· 1
      1.1.1 The current application of smart clothing ······································ 2
      1.1.2 Definition of interactive clothing ·· 4
      1.1.3 Research ideas and methods ··· 6
      1.1.4 Academic and industry research background ································ 7
  1.2 Existing problems and research motivations ·· 13
  1.3 Research objectives ··· 16
  1.4 Research question ·· 17
      1.4.1 Major research question ·· 17
      1.4.2 Subsidiary research questions ·· 18
  1.5 Significance of this book ·· 18

Chapter 2
Literature Review, Definition and Deduction ··· 21
  2.1 The social psychology and semiotic of clothing ······································ 21
      2.1.1 Symbolic semantic of clothing ·· 21
      2.1.2 The inference and representation symbols of clothing ················ 22
  2.2 Maslow's needs psychology and clothing ··· 23
      2.2.1 Maslow's hierarchy of needs ··· 23
      2.2.2 Enlightenment of Maslow's theory on clothing demand ············· 25
      2.2.3 The relationship between Maslow's hierarchy of needs and clothing needs ·· 27
  2.3 The hierarchy and typology of clothing design ······································· 28
  2.4 Deduce the evolution trend of interactive clothing—IoC ························ 30
  2.5 The relationship between DIKWS and the evolution of clothing function ··· 33

2.6　The relationship between CPS/IoT and clothing ·············· 35
2.7　Other associated theoretical research ·························· 37
　　2.7.1　Technology-centered design and human-centered design ·············· 37
　　2.7.2　Emotional design ·············· 37
　　2.7.3　Kansei Engineering ·············· 38
2.8　Summary ·············· 38
　　2.8.1　Interactive clothing conforms to the evolution trend of clothing ······ 39
　　2.8.2　Humanistic and technical features of interactive clothing ·············· 39

Chapter 3
Research Model and Methodologies ·············· 43
3.1　Research roadmap ·············· 43
3.2　Creating the CPCS research model ·············· 44
　　3.2.1　Physical level: entity item of clothing ·············· 44
　　3.2.2　Cyber level: spatial-temporal concept of clothing ·············· 46
　　3.2.3　Social level: humanities activities with interactive clothing as the medium ·············· 46
　　3.2.4　Cross-level: transform from cyber to social structure ·············· 47
3.3　The definition of research methods ·············· 47
　　3.3.1　Practice validation and prototyping methods ·············· 47
　　3.3.2　Prototype evaluation method ·············· 47
　　3.3.3　Model inductive method ·············· 48
3.4　Summary ·············· 48

Chapter 4
Prototyping of Interactive Clothing ·············· 51
4.1　Case study 1, smart infant clothing ·············· 51
　　4.1.1　Infant's crotch humidity monitoring alarm trousers ·············· 51
　　4.1.2　Temperature monitoring infant clothing ·············· 53
4.2　Case study 2, interactive couple clothing ·············· 54
　　4.2.1　Determining the categories of experiential prototype ·············· 55
　　4.2.2　Purpose, methodology and findings ·············· 56
　　4.2.3　Hypotheses and methodology ·············· 56
　　4.2.4　Experimental approach ·············· 57
　　4.2.5　Results ·············· 60
4.3　Case study 3, interactive parent-child clothing ·············· 60

4.3.1 Purpose, methodology and findings ······ 60
4.3.2 Background and research question ······ 61
4.3.3 Prototype design following the CPCS model ······ 62
4.4 Summary ······ 67

Chapter 5
Emotional Evaluation, Data Analysis, and Discussion ······ 69

5.1 Data analysis and evaluation on interactive couple clothing ······ 69
    5.1.1 Kansei evaluation Ⅰ ······ 69
    5.1.2 Kansei evaluation Ⅱ ······ 70
    5.1.3 Case discussion ······ 74
    5.1.4 Limitations and further research ······ 76
5.2 Data analysis and evaluation on interactive parent-child clothing ······ 76
    5.2.1 Selection of comparative evaluation objects ······ 76
    5.2.2 Selection of semantic opposite adjective ······ 78
    5.2.3 Participants' selection and evaluation methods ······ 79
    5.2.4 Differences between the two categories of parent-child clothing ······ 81
    5.2.5 Limitations and follow-up studies ······ 83
    5.2.6 Case conclusion ······ 84
5.3 Summary ······ 84

Chapter 6
Case Study: Purchase Intention and Design Elements of Parent-child Clothing ······ 87

6.1 Case introduction ······ 87
6.2 Research model and hypotheses ······ 87
    6.2.1 PU ······ 88
    6.2.2 PE ······ 89
    6.2.3 PR ······ 89
    6.2.4 FUN ······ 89
    6.2.5 COM ······ 90
    6.2.6 AES ······ 90
    6.2.7 ATT and PI ······ 91
6.3 Method ······ 91
    6.3.1 Sample and procedure ······ 91
    6.3.2 Instrument development ······ 91

6.4	Measurement model	92
	6.4.1 Common factor extraction and naming	92
	6.4.2 Measurement validity and reliability	93
6.5	Hypotheses testing and inspiration	95
6.6	Summary	97

Chapter 7
Knowledge Transfer and Enlightenment to Interactive Clothing Design — 101

7.1	Knowledge transfer	101
7.2	Completion of loops steps in the CPCS model	102
7.3	The "C" design principle for interactive clothing	105
7.4	The framework of the evaluation criteria for interactive clothing design	109
	7.4.1 Background and significance	109
	7.4.2 The structure of the evaluation criteria	109

Chapter 8
Conclusion — 113

8.1	Contribution	113
	8.1.1 Main practical contribution of this book	113
	8.1.2 Original contribution to knowledge science	115
8.2	Future research	116

Bibliography — 119

Chapter 1
Introduction

1.1 Research background

Technology makes fashion more avant-garde, fashion makes science and technology sexier. The internet technology has continuously been integrated into social life, encompassing all aspects of clothing, food, and living conditions, and also changing our production methods and lifestyles profoundly.

Interpersonal communication is a kind of social relations, which is the main content of human life activities. Apart from using language as a tool, human interaction needs to be expressed by some non-linguistic tools, and clothing is one of the most critical non-linguistic tools. With the development of society, clothing plays a distinct role in human interactive activities.

As daily necessities of human life, clothing, especially fashion clothing, carries increasing physical and social properties. The development of clothing can also reflect the progress of science and technology, and often create a stunning topic such as wearable devices or smart clothing that has been moved from the science fiction movie to the present market. This emerging field of smart clothing has brought a dramatic impact on academic research.

Since Thorp (1966) mentioned a wearable computer idea in 1955 and completed the device design in 1961 in his book *Beat Dealer*, and later Bass (1985) described shoe-based computers of the 1970s designed, wearable devices have developed rapidly. As early as a decade ago, researchers began to systematically explore the main development direction of smart clothing in the future (Ariyatum et al., 2005; Cho et al., 2009; Suh et al., 2010; Tao, 2001). In recent years, researchers have generally focused on high-tech approaches to implement smart clothing design, with fruitful success in multifunction (Bahadir et al., 2013; Kan et al., 2015; Schüll, 2016; Perovich et al., 2014; Wright and Keith, 2014; Yu et al., 2014).

In recent years, the focus of Cyber-Physical Systems (CPS) and the Internet of Things (IoT) have moved looks to the future of technology and how it can enhance human activity and experience. As a multi-dimensional smart mechanism with a deep interaction between the physical and cyber world, the CPS/IoT will provide the foundation of emerging and future smart services, and improve our quality of life in many areas (Duarte Filho et al., 2015; Huang et al., 2018; Barnaghi et al., 2015). Following the ascending application of CPS to the fashion industry, the academic and industry research concerns in this area across various disciplines have mainly been devoted to optimizing the interaction of smart clothing through the integration of information technology. The remarkable functions of carefully crafted are generating considerable interests (Barfield, 2015; Gilsoo, 2010), the development of smart clothing in daily life has aroused more attention, and been adopted as the target direction of this research project.

- **Concept of the CPS**

CPS is usually defined as a tight integration of computation, communication, and control with a deep interaction between physical and cyber elements in which embedded devices, such as different sensors and actuators, are wireless or wired networked to sense, monitor, and control the physical world (Lee, 2010, 2015).

- **Concept of the IoT**

In addition to the introduction in 2.4, the IoT can be perceived as a far-reaching vision with technological and societal implications. From the perspective of technical standardization, the IoT can be viewed as a global infrastructure for the information society, enabling advanced services by interconnecting (physical and virtual) things based on existing and evolving interoperable information and communication technologies.

According to the definition of the International Telecommunication Union (ITU, Y. 2060), the IoT mainly solves the interconnection between Thing to Thing (T2T), Human to Thing (H2T), and Human to Human (H2H). Essentially, the interaction between human and machine, machine and machine is mostly to achieve the interaction of information between humans.

1.1.1 The current application of smart clothing

The content covered in this book is interactive clothing, a topic on which there is little previous research, a category of smart clothing in the field of information science, which emphasizes the interaction between humans and the environment based on the integration of information science and traditional clothing, rather than the biological material science category of Smart bio-sensing Clothing.

At present, the research and development of smart clothing has two main ways:

On the one hand, the clothing materials are improved using chemical and physical methods from the "intelligent fiber" which refers to the changes in fiber length, shape, color, temperature, etc, due to the stimulated or change of external environments, such as color-changing fiber, shape memory fiber, tempering fiber, etc. (Tao, 2001, Lv, & Chen, 2016). The typical research projects include Suzanne Lee's "BioCouture" (2005) (Figure 1-1) that the clothing was "grown" from plant cells, and "Biohybrid wearable" with breathing ability which was designed by Yao, et al. (2015) and Wang, et al. (2017) (Figure 1-2).

Figure 1-1 "BioCouture" projects (Top 50 Best Innovations of 2010 in *Time* magazine's annual roundup)
Source: The campus network of Carnegie Mellon University

Figure 1-2 "Biohybrid wearable" projects
Source: Official website of MIT Media Lab

On the other hand, "smart clothing", some of the more advanced examples can be seen in Figure 1-3, which embedded with electronic devices are the combination of electronic devices such as sensors, actuators, memory, data processors, and communication components with clothing by means of embedding (Cho, 2009 a, b; Honarvar, & Latifi, 2016). Electronic products are the scientific and technological basis of clothing intellectualization, and at the same time, clothing has become the carrier of electronic product humanization. The integration of technology and fashion can be realized by embedding electronic devices into clothing, which not only does not affect the wearing effect, but also gives clothing information perception, computing, communication, control, and other capabilities (McCann, & Bryson, 2009; Tao, 2005; Sazonov, & Neuman, 2014, Persson, et al., 2018). Interactive clothing is one of the second categories of smart clothing.

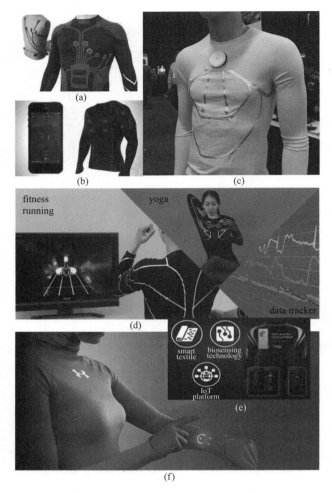

Figure 1-3　Smart clothing examples

Source: (a) Morrison, T. , Silver, J. , & Otis, B. (2014, June). A single-chip encrypted wireless 12-lead ECG smart shirt for continuous health monitoring. In 2014 Symposium on VLSI Circuits Digest of Technical Papers, (pp. 1-2). IEEE.
(b) Official website of Wearable Technologies AG
(c) Official website of Engadget
(d) Official website of Kickstarter
(e) Official website of Slideshare
(f) Official website of Systweak

1.1.2　Definition of interactive clothing

Interactive clothing is an evolutionary branch of smart clothing, and also originates from the wearable devices or wearable computing (Figure 1-4). Gepperth (2012) separated the application scenarios of wearable computing into three categories: "Sensing and Data Analysis", "Interfaces" and "Functionality and Aesthetics". Ariyatum (2005), Van Langenhove & Hertleer (2003) and Zhang & Tao (2001) have summarized the type of interactivity of smart clothing and divided it into three categories: the first is only sense stimuli from the environment, it is one-way communication; the second can sense and react accordingly, it is two-way communication; and the third is a complete interactive activity that it can sense, react and adapt themselves to the environmental conditions, it is interactive communication. At present, smart clothing has realized the first two communication functions under the concept of the IoT, and interactive clothing is precisely born in response to the third communication function.

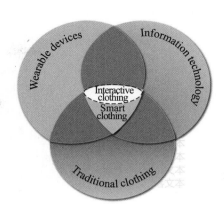

Figure 1-4 The relationship between wearable devices, information technology, traditional clothing, smart clothing, and interactive clothing

Although there is no clear academic definition of interactive clothing so far, interactive clothing should include the following characteristics:

- Interaction occurs as two or more clothing have a two-way effect upon one another. In the aspect of interactivity, it refers to the interaction between clothing and the body of the wearer, clothing interacts with the environment, and clothing interacts with other dressers. The interactive clothing can sense, react, and adapt themselves to the environmental conditions, and it is an interactive communication.
- Covering a wide range of interdisciplinary, including clothing design and engineering, information science, human-computer interaction, textile science, social psychology and industrial & art design. As Gepperth emphasized, aesthetics is an essential element.
- In essence, it is still clothing, and the interactivity is an additional feature. No matter how sophisticated and changeable the interaction is, it should not be divorced from the essential attributes and basic functions of clothing.
- In addition to the special information science function of Self-Monitoring Analysis and Reporting Technology (S.M.A.R.T., the term derives from the data security technologies commonly used in hard drives) that are owned by smart clothing and basic physical functions of traditional clothing, the interactive clothing is more emphasizes that clothing can carry the media function of society and humanities in the process of interactive activities between humans and their environment based on information technology.
- Combining CPS with clothing engineering design to input a certain physical signal into clothing, so the clothing can output a specific social symbol that people or the clothing environment can perceive and generate corresponding interaction. This kind of social symbol is the expression form of clothing as the interactive medium for people to interact with the environment.

1.1.3 Research ideas and methods

The content covered in this book was carried out according to the following logical sequence (Figure 1-5):

- The research stage planning mainly includes the theoretical analysis stage, model building stage, practical application stage, model verification, and summary stage.
- Regarding research contents planning, it contains mostly the derivation and building of the theoretical model, prototype development and emotional evaluation, and model verification.
- Concerning research methodology planning, it mainly includes a literature review, questionnaire, prototyping, Kansei Engineering, Semantic Differential (SD) method, and so on.

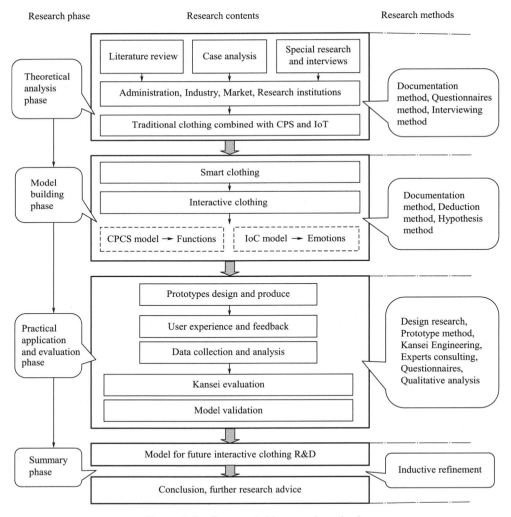

Figure 1-5　Research ideas and methods

1.1.4 Academic and industry research background

Because interactive clothing originates from smart clothing, the related research process analysis should be based on the development process of smart clothing and wearable devices.

A literature review was carried out to investigate the R&D trend of smart clothing and wearable devices. Wearable devices have a long history of development, ideas and prototypes emerged in the 1960s, while devices with wearability appeared in the 1970s and 1980s. The wearable computer prototype developed by MIT Professor Steve Mann is one of the representatives. With the rapid development of computer standardized software and hardware as well as internet technology, wearable smart devices have become diversified, which has gradually shown significant research value and application potential in many fields such as industry, medicine, military, education, entertainment, and so on.

The methods and approaches of the literature review are as follows:

Firstly, the number of patent applications related to wearable devices is taken as the basis for analyzing the development process of smart clothing (Yan et al., 2016). The Derwent Innovation Index database was taken as a patent retrieval platform, a retrieval strategy with the theme of "wearable devices" constructed by technology decomposition. The results are de-noising, and the artificial interpretation determines the analytical data range. Figure 1-6 shows the annual trend in the number of patent applications related to wearable devices from 1974 to 2018 May and also can be seen that the three years with the highest number of annual patent applications were 8016 records in 2014, followed by 5066 records in 2013 and 4898 records in 2015 (Figure 1-7).

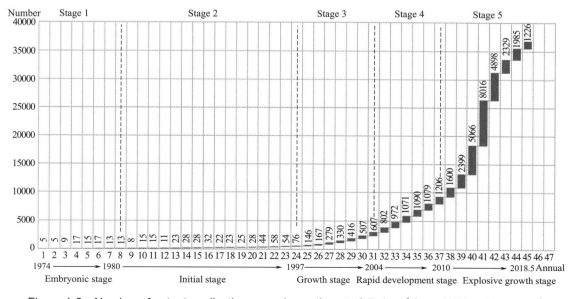

Figure 1-6 Number of patent application annual growth waterfall chart (from 1974 to 2018 May)

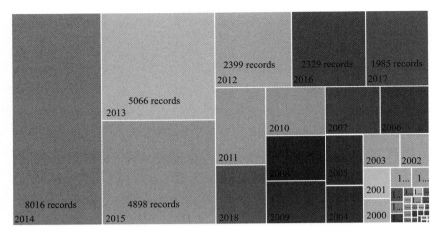

Figure 1-7   Annual number of the patent application treemap

Secondly, the number of research papers related to smart clothing is taken as another basis for analyzing the development of smart clothing. The Web of Science was selected for the research papers retrieval platform, with all databases as the target database, smart clothing as the keyword of the search topic, the deadline is October 2018. Figure 1-8 shows the trend in the number of research papers on smart clothing from 1983 to 2018, a total of 811 records. The source Databases includes Web of Science Core Collection (595 records), MEDLINE® (252 records), KCI-Korean Journal Database (171 papers).

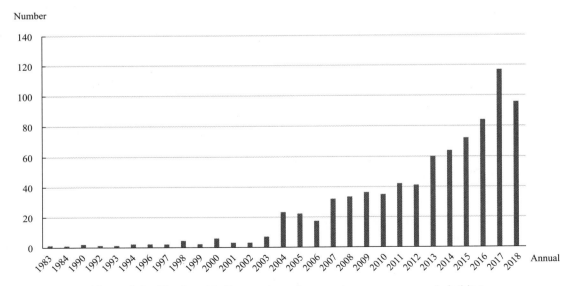

Figure 1-8   The trend in the number of research papers on smart clothing

Compared with the number of patents, research projects, and academic journals (Amft, & Lukowicz, 2009; Ariyatum, & Holland, 2003; Ariyatum, et al., 2005), the results show that the patented technology in wearable devices mainly focuses on the data input/output devices, diagnostic measurement, and human body identification, data processing, communication devices, wireless communications, optical systems, data security protection, program control, graphics recognition, sensors and other aspects (Table 1-1). The design evolution of smart clothing could be divided into five stages (Labeled in Figure 1-6) as follows:

Table 1-1 Classification of wearable devices research area

Research Area	Product Category
Fitness Health	Sports monitors, fitness and heart rate monitors, smart sports glasses, smart clothing, sleep sensors, and mood measurements
Healthcare	Cardiac and insomnia monitoring, stress detection, diabetic nephropathy treatment, wound infection Detection and treatment, surgical navigation, and functional rehabilitation training
Industrial Military	Driving safety monitoring, navigation, military exoskeleton, visual helmet night-vision goggles, and smart bulletproof vests
Infotainment	Smart watches, smart glasses, augmented reality headphones, virtual reality helmet, and mind headband

(1) **The embryonic stage (1960-1979).**

Edward O. Thorp, a professor of mathematics at MIT, mentioned in his book *Beat Dealer* (1966) that he first thought of a wearable computer idea in 1955 to increase the winning percentage of roulette. In 1960-1961, he was in cooperation with another inventor to complete the development of the device, which could represent the original prototype of wearable devices.

By the 1970s, some institutions in the United States and Japan had begun to apply for relevant patents in the field of wearable technology. The number of patents in the initial stage was minimal, with an annual average of fewer than 50 patents. In 1977, C. C. Colins, from Smith Kettlewell Eye Research Institute, designed a vest for the blind or visually impaired, whose head-mounted camera was converted into tactile imagery through the mesh on the vest, allowing the blind to "see" (Swanson, 2014).

(2) **The initial stage (1980-1997).**

During this period, design methods were considered as technology-driven (Orth, 1998) because most research and development focused on wearable devices and the application of advanced technologies. For example, Steve Mann (1996, 1997) embedded controllable camera equipment in a specially designed backpack with a head-mounted camera. Randell (2001) predicted that the integration of sensing and display technologies would create more opportunities for textile manufacturers. So the researchers and their team developed Cyber Jacket, which integrates position sensors (GPS), displays, and so on, to prove this possibility. However, this prediction was based on the trend towards the miniaturization of electronic devices, not consumer demands. Whether or not people carry electronic devices is controversial, which means they want them to be part of their clothing and operate unnoticed without user awareness. Besides, clothing design and business input have been neglected. As a result, products are more "portable" than "wearable", for example, the MIT wearable computer (O'Mahony, 2002).

(3) The growth stage (1998-2004).

From the end of the 20th century to the beginning of the 21st century, the internet was booming. During this period, the microprocessor, sensors, and other electronic components of the production of small volume, low power consumption, and high precision continuous development. With the constant improvement of computer function, the quality of image processing is getting higher. These technologies have led to the development of wearable devices, and the number of patent applications in this field has begun to increase significantly and grew faster.

The awareness and participation of the fashion and textile industry increased significantly in these four stage types of organizations involved in smart clothing development as shown in Table 1-2.

Table 1-2 Categories of organizations in 1998−2004

No.	Categories	Organizations
1	Academic institutes	MIT Media Lab, Royal College of Art, University of Bristol, Central Saint Martins College of Art & Design
2	Governmental organization	NASA in the USA and Ministry of Defence in the UK
3	High-tech companies	Nokia, Philips, Ericsson, Motorola
4	Clothing companies	Levi's, Polo

As a result, the number of collaborative projects in the electronics and fashion sectors has increased rapidly, such as the Cyberia project (Rantanen et al., 2000). Also, textile and clothing experts began to create their R&D (Clarke and O'Mahony, 1998). Senior fashion designer Alexandra Fede, for example, has worked with DuPont to develop new smart clothing by advanced technology. Although applications have become more wear-resistant, most of the results are still prototype clothing because of technological underdevelopment, such as Philips and Levi's ICD+ jackets (Meoli and May-Plumlee, 2002) and SCOTT eVast (Forman, 2001) although they have offered in limited numbers and focused on the niche market. In addition, product concepts, such as embedding fiber-optic screens on clothing (Gould, 2003), still do not satisfy the requirements of the mass market.

The number of smart clothing available on the market increased. Notable examples are the Met 5^{TM} jacket (Ward, 2001), Adidas Smart Shoes, and GapKid sweaters with embedded FM radios. Most development teams (Rantanen, 2000) have adopted a wide range of multidisciplinary approaches and user-centric design. In addition, the boundaries of applications extend to new areas.

It also can be seen from Table 1-2 that the electronics sector is more actively involved in project development than fashion because most projects were led by the electronics R&D team during this period.

(4) **The rapid development stage (2005-2010).**

During this period, smart clothing was no longer embedded with simply electronic components but emphasized the development of smart textile materials. As a result, the number of wearable devices gradually increased, its wearability has been significantly improved.

Professor Takagi (1990) first proposed the concept of intelligent materials based on the idea of integrating information science into the configuration and function of materials. Later, American scholars called it smart material (Gu, & Chen, 2006). Professor Shi Changxu, editor-in-chief of *the Material Dictionary*, explained smart material as a material that mimics the life system with both sensory and driving dual functions. That is, the smart materials cannot only perceive the change in the external environment or internal state, but also can change one or more properties of the materials in real-time, and make the desired response, also known as smart materials, through some feedback mechanism of material itself or outside (Schwartz, 2002). Perception, feedback, and response are the three main elements of smart materials. One notable feature of smart materials is the combination of high-tech sensors and actuators with traditional materials, giving the material new properties that make inorganic materials more and more biologically specific properties. Compared to smart fiber, smart clothing is more accessible to achieve "smart" by integrating sensing elements, feedback elements, and response elements.

The concept of "interactive smart textiles" was put forward by Professor Tao (2006) is one of great practical significance and guidance for smart clothing development. She proposed that interactive smart textiles refer to textiles that can perceive and respond to stimuli or environmental conditions dominated by electricity and light in an artificial or predetermined manner. She also contributed to wearable electronics and photonics (2005) which covers topics related to materials and devices, structures and system architectures, and various issues that fashion designers need to face.

(5) **The explosive growth stage (2011 to present).**

Since 2011, wearable technology patent applications have shown a rapid development trend, more and more fashion companies and institutions have been involved in this field, and continue to develop new wearable products, the number of patent applications has become an explosive growth trend. This phase was characterized by institutions taking market share firstly through new products and then planned their patent layout in various ways (Wei, 2014). Internet companies continue to expand their scope of technology protection through the method of patent acquisition. With the accumulation of technology for many years, mobile terminal enterprises continue to set up new barriers to technology protection. However, with the increasing improvement of the CPS and the IoT technology in the information technology industry, wearable does not only meet people's entertainment and leisure needs but also develops towards health monitoring, convenient medical care, and intelligent manufacturing.

Contrary to the growth stage (1998-2004), since 2011, the enthusiasm of the fashion industry to participate in smart apparel is significantly higher than the information electronics industry, and smart clothing has begun to shift from focusing on function in the past to daily life clothing. Although the smart clothing industry is still in its infancy, it is characterized by fierce competition and a large number of new entrants. Some of the most famous companies in the smart clothing market are shown in Table 1-3.

Table 1-3 Famous companies involved in the competition of the smart clothing market

Company	Location	Domain
AIQ Smart Clothing Inc.	Taiwan, China	Sports & fitness, home & leisure, outdoor & leisure, home care & health care
Clothing Plus Ltd.	Finland; USA	Fitness and Healthcare
Textronics, Inc. (Adidas)	USA	Sports & fitness, health & wellness, military & safety
Sensium Healthcare	UK	Healthcare
VivoMetrics	USA	Wearable technology
Schoeller Technologies AG	North America; Europe; Asia	Functional clothing and textiles
Grado Zero Espace	Italy, Europe	Wearable Technology
Athos	California, USA	Fitness & Healthcare
Catapult Sports	Australia	Wearable Technology
Heddoko	Montréal, Canada	Ergonomics and Sports
Hexoskin	Montréal, Canada	Fitness & Healthcare
OMsignal	Montréal, Canada	Wearable Technology
Sensoria Inc.	USA	Fitness & Healthcare
Ralph Lauren Corporation	USA	Sports & Fitness
LikeAGlove	Israel	Wearable Technology

Source: Hanuska, et al. (2016). *Smart clothing market analysis*. University of California Berkeley.

According to the ABI Research's Wearable Data Analytics and Business Models report released in Dec. 2017, "Wearable data and analytics services revenue will reach over US$838 million in 2022, increasing from over US$247 million in 2017, a CAGR of over 27%; The smart clothing market will get a 45% CAGR". In recent years, researchers have generally focused on CPS/IoT approaches to implementing smart clothing design, with fruitful success in sports, health care, and other fields (Bahadir et al., 2013; Kan et al., 2015; Perovich et al., 2014), and a small amount of sportswear has been put into the market. The 2016 and 2017 Milan, Paris, New York, and Tokyo Fashion Week shows, especially the Fashion Tech Festival 2017 in Paris, 2018 and 2019 in Berlin, which is bringing together fashion designers, engineers, and major IT companies invested in developing the future of design for smart clothing and accessories, more and more fashion brands' product development and the effectiveness of the fashion show have adopted CPS/IoT approaches.

Nevertheless, some products are just accessories of clothing rather than the clothing itself. Some intelligent effects are limited to specific places. Undeniably, technological innovation can enhance the fashion brand's marketing campaign and catch consumers' great attention, CPS will be integrated into our daily clothing life in the future (Wright and Keith, 2014; Schüll, 2016).

1.2 Existing problems and research motivations

Despite its impressive track record in smart clothing research and development, we still have to confront with several existing research problems or development bottlenecks from correlative humanities science and information technology respectively. These existing problems and challenges are the motivation of this research to choose an interactive clothing project as the research topic.

(1) From the perspective of social psychology and human-centric design principles, the smart clothing R&D approach is not driving on the right path.

According to the perspective of social psychology, human beings are living in human society, interpersonal communication, as well as individual symbolic interaction, are the foundation for individuals to form the activities of society (Blunter, 1994). Clothing that people wear should be an auxiliary sign or code means for people to interact in society (Leeds-Hurwitz, 2012). Therefore, according to the human-centric product development principles (Buchanan, 2001; Norman, 2005; Steen, 2012), smart clothing should be extended to the direction of assisting interpersonal communication in social activities. This assisting interpersonal communication is precisely the value and significance of interactive clothing research.

Blunter (1994) has proposed that human society is a symbolic interaction. He defined "symbolic interaction" as the distinctive character of communication as it takes place between human beings.

"The peculiarity consists in the fact that human beings interpret or define each other's actions instead of merely reacting to each other's actions. Their response is not made directly to the actions of one another but instead is based on the meaning which they attach to such actions. Thus, human interaction is mediated by the use of symbols, by interpretation, or by ascertaining the meaning of one another's actions…" (Blunter, 1994, pp. 263)

"Human society is made up of individuals who have selves; that individual action is a construction, being built up by the individual through noting and interpreting features of the situations in which he acts; that group or collective action consists of the aligning of individual actions, brought about by the individuals interpreting or taking into account each other's actions...." (Blunter, 1994, pp. 266)

(2) From the perspective of clothing design, smart clothing lacks aesthetic purposes or the humanistic attributes of fashion.

The imbalanced contribution from electronics and fashion industries is undeniable (Wang et al., 2018). According to the research papers retrieval and results analysis on the Web of Science, as of October 2018, the top 25 research fields of 812 research papers related to smart clothing (Figure 1-9), excluding the field of humanities and art design. It is a shocking result. It can be concluded that long-term research on smart clothing has only experienced technological development without artistic design rendering.

The development of smart clothing is not yet known as clothing design in the field of humanities and art, and it is not fashionable enough to enter our daily wardrobe. According

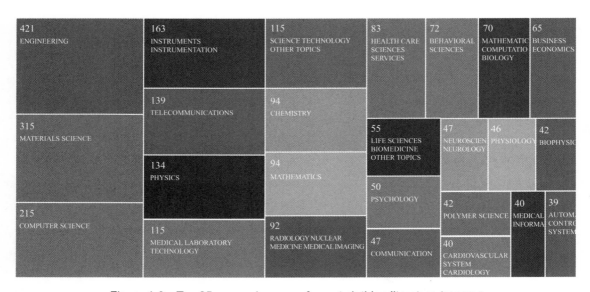

Figure 1-9　Top 25 research areas of smart clothing literature treemap

to the five stages in the development of wearable devices mentioned above, rather than fashion designers, IT and electronic technologists are leading this emerging area (Stead et al., 2004).

(3) From the perspective of the material and social attributes of clothing, some smart clothing has become the dominant of the wearer rather than the wearer's appendage.

Because many innovations in this field are driven by technological development, rather than users' needs and preferences (Steen, 2012), the result of putting the cart before the horse leads to the smart clothing not serving people, but people serving the clothing. Clothing should be an appendage of human beings, and should not make people an accessory to clothing.

(4) The contradiction between clothing as a high-tech carrier and clothing as an interactive symbol of human emotion carrier.

Researchers have generally focused on high-tech approaches to implementing smart clothing design with multifunction. Nevertheless, the complex sociological aspects of clothing, i.e., its interactive symbolism and properties of emotional expression, should not be ignored. A long time ago, Naisbitt & Cracknell (1982) had proposed that we are moving toward "High Tech" and "High Touch" in two directions and humans are trying to give each new technology a compensatory response. This reaction is a kind of non-material emotional value, humanistic connotation pursuit, and the balance of design and emotional factors. Clothing has both material attributes, and complex social attributes (Barnard, 2002; Owyong, 2009; Weizhen et al., 2017), the expression function of emotion is an important design element which cannot be neglected in the development of smart clothing.

(5) In the perspective of information technology, the architecture model of smart clothing development is not perfect.

When we extend the functional definition of interactive clothing to provide interaction between different wearers or groups rather than just personal interaction with data, it is a significant challenge to apply Cyber-physical systems (CPS) in the development of interactive clothing. That is, how to meet the differentiated needs of different wearers and the differences in various attributes among each wear groups. Also, most of the existing CPS architecture research is based on the logic framework of information science, but not the specific technical guidance for the non-information engineering personnel. Therefore, it is critical to optimize the CPS technical architecture which should comply with wide applicability for interactive clothing (Chen et al., 2016; Alhafidh et al., 2017; Longo et al., 2017).

(6) From the perspective of network communication under the IoT architecture to assess, the communication of current smart clothing is not perfect enough.

Rafaeli, Sudweeks & McLaughlin (1998) proposed that the net communication is

15

divided into three levels: one-way communication, two-way communication, and interactive communication. Ariyatum (2005), Van Langenhove & Hertleer (2003) and Zhang & Tao (2001) have summarized the type of interactivity of smart clothing and divided it into three categories: the first is only to sense stimuli from the environment; the second is to sense and react accordingly; the third is a fully interactive activity that it can detect, react and adapt themselves to the environmental conditions. The current smart clothing only implements the first two functions, and the third function requires a new type of clothing to undertake this historic mission.

(7) Failure to systematically interpret the principles and processes of interaction.

Current researchers seldom dabble in the systematic research of smart clothing from information modelling to humanistic evaluation, most of them analyzed the information technology application or humanistic design from the single academic background (Trindade et al., 2016; Kim et al., 2016b; Salim et al., 2014; Weizhen et al., 2017), which may lead their inability to systematically explain the principle and process of how to extract data, transform and form humanistic feelings of the interactive clothing.

(8) The current category of smart clothing is not rich enough to fit into daily life, limited to the stage effect of the fashion shows.

Most smart clothing development is applied to particular functions or special situations rather than to the needs of everyday life. It has been suggested in the previous study (Wang et al. 2018) that some fashion brands have adopted VR and other technologies in the fashion show rendering the atmosphere, to form the interaction between models and the audience. However, the visual effects of these clothing will disappear if leave the show scene. Obviously, this design has not yet entered the daily clothing life category.

As Prof. Hiroshi Ishii from MIT proposed at the 2016 Symposium on System Integration, the technology soon becomes obsolete, but the true research visions can last longer. For smart clothing or interactive clothing research, art & philosophy, design & technology, art & aesthetics should all be coordinated development. The choice of interactive clothing as a research topic is to deal with the current situation and challenges of the seven deficiencies mention above, and strive to develop a new approach for the research of interactive clothing from the perspective of academic theory and practice.

1.3 Research objectives

The overall purposes of this book are to bridge the gap between CPS in the field of information science and emotional evaluation in the field of humanistic science, to minimize the unbalance between human emotions and wearable smart technologies for interactive clothing innovation, to investigate how the transformation could be realized from information to knowledge in the process of interactive clothing design. By optimizing the CPS architecture model of interactive clothing design and conducting the Kansei Engineering (Nagamachi, 2011a, b) analysis of wearers' evaluation, the R&D elements based on the CPS or the IoT technology were explored, and practical implication for interactive clothing design was suggested.

Specific research objectives were as follows:

To verify the concept of smart clothing development that "technology makes fashion more avant-garde, fashion makes science and technology sexier". To break the undesirable situation that the current research of smart clothing concentrates on health care or sport monitoring fields, but ignores the demand of the social interpersonal communication in wearer's daily life.

- To define the concept of interactive clothing, build an R&D theoretical model for interactive clothing.
- To build an innovative CPCS model, analyse its feasibility with the interactive clothing prototyping development as a case.
- To explore the development direction of smart clothing in the future and analyze the possibility of realizing the Internet of Clothes (IoC) based on CPS/IoT technology.
- To investigate the transformation from the input of the physical signal to the output of the social symbol in the prototype development process.
- To establish a basic framework of interactive clothing design evaluation criteria, encourage the object of evaluation to develop in the right direction and objectives.
- To reveal the design and development principles of interactive clothing from the perspective of humanities and technology, guide the development of interactive clothing.

1.4 Research question

To attain the objectives mentioned above, this book will answer one major research question (MRQ) and three subsidiary research questions (SRQ).

1.4.1 Major research question

What happened when the social semantic of the interactive clothing transformed from information stage to knowledge stage in the architecture of CPS?

This question includes three levels of content:
- According to the research dimensions of clothing social psychology and semiotics, as a social product, clothing has special functions of social symbols and information transmission (Steffen, 2009a). Since the dimensions of communication and characterization are more evident in clothing than those of most other commodities, this

field has been attracted to research topics from the perspective of anthropology (Sahlins, 1976), psychology (Sommer, & Wind, 1986), sociology (König, 1971) and semiotics (Barthes, 1983). The research of interactive clothing is also inseparable from these fields.

- Although interactive clothing is advertised as a scientific and technological product, it is still clothing, in essence. Therefore, the research of interactive clothing also needs to explore its smart semantic (Steffen et al., 2009b) such as information transmission function.

- The smart semantic transformation of interactive clothing is also the realization process of data-information-knowledge-wisdom (Krippendorff, 2006). How to apply the CPS principle of information science to realize the change from physical information into a kind of consensus semantic of human society is an interdisciplinary research question.

1.4.2 Subsidiary research questions

Three subsidiary research questions (SRQ), according to soft system methodology, i. e., the "WHY," "HOW" and "WHAT" as follows.

SRQ 1: Exploration of research and development approaches.

Why does the R&D of interactive clothing worn in daily life need to integrate smart technology with emotional design and humanistic evaluation simultaneously?

SRQ 2: Investigating technical means.

How could technology transformation be realized between the signals in the field of information physics and the symbols in the various dimensions of humanity and society?

SRQ 3: Refinement of design principles.

What does the design principles of interactive clothing should be valued in the context of CPS's growing prosperity?

1.5 Significance of this book

The research topic of this book belongs to interdisciplinary research, a topic on which there is little previous research. The subject takes interactive clothing as the research carrier, integrates data extraction and signal input in the field of information science, with humanistic emotional evaluation and the output of social symbols in the field of knowledge science, and conducts theoretical and practical research from the perspective of application in the CPS/IoT era.

(1) **Theoretical and academic level.**

- Conceptual innovation.

This research innovatively and clearly puts forward the concept of interactive clothing, a topic on which there is little previous research, will also explore the new development path of smart clothing.

- The significance to knowledge science.

This research will integrate the human and technical perspective, put forward the technical development process, design principles and evaluation criteria framework of interactive clothing, and initially establish the knowledge framework of interactive clothing design and evaluation.

The level and function of the interaction between human beings and the environment, human interaction, object and object interaction in knowledge science will be identified and classified. Taking the prototype development process of interactive clothing as case, the DIKW model from data to information, from information to knowledge, from knowledge to wisdom will be systematically completed. From the perspective of knowledge management, this research will dismantle and analyse the transformation from the input of the physical signal to the output of the social symbol in the prototype development process of interactive clothing.

(2) **Industrial level.**

This topic expands the industry chain of the clothing industry, not only integrating the main elements of the traditional clothing industry but also adding CPS/IoT technology, which makes clothing and clothing, clothing and environment, clothing and people interact. More importantly, it puts forward the concept of the IoC, that is, the industrial prospect of interaction between people in social groups through clothing.

(3) **Application level.**

This topic focused on the social needs of the next 10-20 years, especially the interactive clothing needs of the ageing society, is of great significance. With the continuous development of the information society, Smart Home, Smart City, and Intelligent Transportation will bring more convenience to our lives. At the same time, in the face of the elderly's living needs, emotional needs and communication needs, the interactive clothing will become a vital carrier medium in the future.

Also, this topic introduces the CPS technology into the design of interactive clothing in the art & humanities field and puts forward the technical architecture of interactive clothing development, that is, an innovative CPCS model, from the perspective of expanding the application field of information technology.

(4) **Product design and development method level.**

In response to the problem of "cold" and lack of emotional semantic in science and technology, this research will integrate human-centered design principle, emotional design methods, and Kansei Engineering Evaluation approach into the field of interactive clothing project development. And advocating the essential attribute of interactive clothing design still belongs to the category of humanities and arts, even if scientific and technological functions accompany this kind of clothing.

Chapter 2
Literature Review, Definition and Deduction

The research on product development or the deduction of its future development cannot be separated from the product's past and current essential attribute and its development law. From the material level to the spiritual level of technical and humanistic perspectives, the related research literatures are classified for building a pyramid deduction analysis model.

2.1 The social psychology and semiotic of clothing

2.1.1 Symbolic semantic of clothing

The study of the symbolic semantic of clothing is helpful to analyze the essential attributes of clothing. As far back as the 17th century, English philosopher Locke (1690) used the word "semiotica" in the last paragraph of his book *An Essay Concerning Human Understanding*, which means "the theory of symbols." In the middle of the 20th century, great strides were made in the research of clothing symbols.

Although some books about clothing and its functions had emerged in the 19th century, interdisciplinary research book on the psychological or social functions of clothing appeared on the first half of the 20th century, for example, *The Psychology of Clothes* (1930).

The Fashion System (Barthes, 1983) is a sociological classic. Its most significant contribution lies in illustrating the symbolism of fashion, explaining that fashion is used by the media as a communication tool to elevate clothing from the material level to the spiritual level. Barthes divided clothing into three categories. The first category is the material level of clothing, which is well understood. The second category is "represented clothing". If the material level of clothing is its original appearance, is represented by the pictures, language re-translation, processed clothing, is not the same as the original clothing, and the spiritual value of clothing is also added in the process. The third category is used clothing, that is, clothing worn by consumers after purchase. This is given a different meaning, such as personal information and emotion.

Clothing is the product of human civilization and an essential part of human culture. Almost from its origins, clothing combines the practicality of protecting the body, maintaining life, and satisfying the role of self-expression. The former is the material attribute of clothing, the latter is the social attribute of clothing (Wang et al., 2018b). With the progress of the society, clothing constantly adds new social attributes. Clothing cannot only satisfy the desire of self-expression, but different styles and materials of clothing in a specific social system can also represent different identities, status, can reflect the different ideas, attitudes, concepts and so on (Morgado, 1993). In this case, clothing is "something that can represent something" which is a symbol. Therefore, we call clothing a symbol is not based on the point of view of material attributes, in fact, the social attribute of clothing is the necessary condition for clothing to become a symbol.

Undoubtedly, clothing symbols belong to a kind of cultural symbol, which have many common characteristics with architectural symbols, film symbols, theatrical symbols and so on. In addition, clothing can not only be used to cover the body, but also convey meaning. Semiologist Roland Barthes (1964) called clothing a functional symbol which originated from practicality.

Langer, Susanne K. (1953, 1957) divided symbols into inference symbols (i.e., linguistic symbols) and representation symbols (i.e., non-linguistic symbols). The former is used for scientific analysis and exchange of ideas, while the latter is used for artistic insight and expression of emotion. Artistic symbols belong to the representation symbols.

On the one hand, clothing, as a necessity of life, plays a role in protecting the body and maintaining life. It must conform to ergonomic principles and meet the requirements of human comfort. In this respect, clothing symbols have the basic characteristics of inference symbols. But at the same time, clothing, as a product of human social life, is a symbol of the human tendency towards "natural humanization" in the development of human culture. It embodies different social images of human beings in different societies with various forms. From this point of view, clothing symbols have the same characteristics as poetry, music, painting and other artistic symbols. Therefore, we cannot merely judge whether the clothing symbol belongs to the inference symbol or the representative symbol.

2.1.2　The inference and representation symbols of clothing

Summarizing the social semiotics and archaeological research results associated with clothing, it can be concluded that clothing has multiple attributes (Wang et al., 2018b). The interrelationships between clothing attributes and their evolutionary path mainly include five (can also be said to be six) attributes or symbols ranging from low-level to high-level as follows (Figure 2-1).

The inference symbols include:

(1) Improving warmth and comfort (Pedersen, 1923; Toups et al., 2010);

(2) Providing modesty and protection (Dunlap, 1928; Kittler et al., 2003).

The representation symbols include:

(1) Displaying logos (Harms, 1938; Gilligan, 2010), and also reflecting a certain level of technological development (Twigg, 2009; Park et al., 2014);

(2) Conveying certain social and cultural connotations (Harms, 1938; Feinberg et al., 1992; Lennon & Davis, 1989a, b; Roach & Eichler, 1979; Owyong, 2009);

(3) Facilitating communication and expression (Bohn, 2004; Norman, 2004; Kaiser, 1983; Hsu and Burns, 2002; Barnard, 2002; Leeds-Hurwitz, 2012).

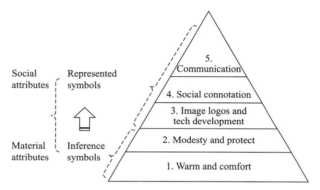

Figure 2-1 Symbolic semantic model of clothing

Through the symbols of clothing, people show the characteristics of their individual or group and realize the identification and connection in communication and interactive activities. Through the interpretation of clothing symbols, people have established a kind of interpersonal communication rules.

Therefore, it is of great significance to interpret human clothing behaviors from the perspective of semiotics and to explore the effect of clothing symbols on interpersonal communication activities or interpersonal relationships. The study of interactive clothing is also inseparable from the interpretation of the semantic of clothing symbols.

2.2 Maslow's needs psychology and clothing

2.2.1 Maslow's hierarchy of needs

Maslow's Hierarchy of Needs (Maslow, 1943; 1968) is a valuable assessment tool that is used in the research fields of clothing demand trend.

As can be seen from Figure 2-2, the five-level hierarchy model can be divided into basic (or deficiency) needs (e. g., physiological, safety, love, and esteem) and growth needs (self-actualization) (McLeod, 2007; Poston, 2009).

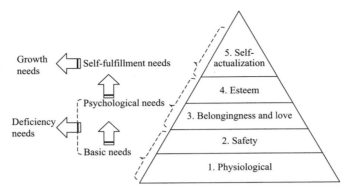

Figure 2-2　Maslow's hierarchy of needs model (Maslow, 1943; McLeod, 2007; Poston, 2009)

Such basic needs can be stimulated when people are not satisfied. Moreover, the need to meet these needs will become stronger with the prolongation of time denying. Individual's needs must satisfy the lower basic requirements before they progress to the higher level of growth needs. Once these needs are reasonably met, individuals are no longer motivated by them, the highest complex level of self-actualization can be achieved (Tay, & Diener, 2011).

The original hierarchy of needs five-level model includes:

(1) **Biological and physiological needs.**

This is the most basic requirement for human survival, such as food, shelter, and clothing. If these needs are not satisfied, human survival will become a problem. In this sense, physiological needs are the most powerful driving force for people to act.

(2) **Safety needs.**

This is the need for human beings to safeguard their safety. Of course, once this demand is relatively satisfied, it will no longer be an incentive factor. The keywords include security, order, law, stability, health and wellness, safety against accidents and injury, freedom from fear.

(3) **Love and belongingness needs.**

People have a feeling of belonging to a group, hope to become a member of the group, and care for each other. The keywords include friendship, intimacy, affection, and love, from workgroup, family, friends, and romantic relationships.

(4) **Esteem needs.**

Everyone wants to have a stable social status and demands that the abilities and achievements of the individual be recognized by society. Maslow believes that respect for the need to be satisfied can make people full of confidence in themselves, enthusiasm for society, and experience their value. The keywords include achievement, mastery, independence, status, dominance, prestige, self-respect, and respect from others.

(5) Self-Actualization needs.

This is the highest level of needs, that is, the peak of the pyramid. It refers to the need to achieve personal ideals, aspirations, to maximize the ability of the individual, to complete all things commensurate with their abilities. Maslow suggested that the approach taken to meet the need for self-realization varies from person to person. The need for self-realization is to strive to realize your potential and make yourself increasingly the person you want to be. The key words include realizing personal potential, self-fulfillment, seeking personal growth and peak experiences.

Maslow's hierarchy of needs theory reveals the common law of human behavior and psychological activity. Maslow explores human motivation and studies human behaviors from the perspective of human needs, and points out that human needs are constantly evolving from lower level to a higher level. Therefore, the hierarchy of needs theory has an enlightening effect on research the trend of clothing demand.

2.2.2 Enlightenment of Maslow's theory on clothing demand

In human social life, there may be no more vivid expression of people's value orientation and way of life than the choice of dress. This study analyzes the embodiment of Maslow's hierarchy of needs theory in the trend of clothing demand from five aspects as follows:

(1) First of all, people need to survive, and human needs affect people's behaviors. A person's physiological needs for food or keep thermal comfort are the first to meet the needs of all requirements. For a person who cannot fulfill his physiological needs, clothing is just for them to keep warm and shame. When a person's needs at a certain level are met at a minimum, a higher level of needs will be pursued. So this is the minimum level of demand from the market because consumers only require clothing products with a general function of shading and warming.

(2) If the physiological needs are relatively satisfied, there will be a kind of safety need, which manifested in the hope that people want to live and work in a safe and orderly environment, with stable occupation and life security. If the security needs are not satisfied, people will have a sense of threat and fear. With the gradual rise of demand, clothing has gradually evolved from a warm function to a tool to protect their bodies and ornament. At this time, the clothing market only needs to satisfy part of the consumer market, which has the requirements for "safety". Consumers pay attention to the impact of clothing products on the body, such as ski suits and climbing shoes with special features and defensive functions, which they need.

(3) With the highly developed modern society and economy, people no longer work and live just for food or clothing, they have the free choice of purchasing and the preference

for clothing. The dress of freedom makes the relationship between individual wearing behavior and self-expression more closely related. The purpose of human dressing is usually summarized as protection, etiquette, and decoration, while the purpose of etiquette and decoration is closely associated fashion.

With the rapid development of society and the acceleration of life rhythm, clothing has also played a distinct role in displaying human personality and psychological conditions. People are eager to be recognized by society whenever and wherever they are. If a person's physical and security needs are well met, there will be a need for love, emotion, and belonging, that is, social needs. Social needs are reflected in clothing. In school, students have school uniforms; in the company, employees have work clothing; in the evening party, there are charming and noble evening clothes; when sleep there are pajamas. At this time, the choice of clothing is expressed as a desire to have a harmonious relationship with peers and colleagues or to maintain a friendship, or belong to a group and become one of them. This is the role that clothing plays in social needs. So at this time, apparel companies began to introduce a variety of social functions of clothing to meet the "communication" requirements of the market, such as a variety of uniforms, suits, professional clothing design, these are consumers concerned about whether the product helps to improve their communicative image requirements.

(4) People need self-esteem and the desire to gain respect from others and hope that personal abilities and achievements could be recognized by society. Clothing in the need for people to show, it can be designed to sell their satisfaction, self-confidence to gain the respect of others.

The clothing we choose can project our thought, impressions, and ambitions. For example, business people will bring their confident appearance and feelings with the latest style or most expensive suits for attending important meetings. At this time, clothing can be used to overcome the anxiety about external appearance or abilities, to help them integrate into the environment and social occasions more confident, can also win respect from others. The physical and safety needs of a senior employee who is wearing high-end fashion brand do not inspire her interest in high-end fashion. Her interest in high-end fashion may come from her status as a professional woman, as consumers seeking to respect demand are more concerned about the symbolism of the product.

(5) In modern society, the changing way of lifestyle has become a significant factor affecting the clothing culture, and clothing was once a symbol and logo to distinguish people's social status and rank.

Through a person's clothing, we can "see" their social status, economic status, political tendency, national affiliation, lifestyle and aesthetic taste. So clothing is a visual language of communication, it can tell us what kind of people the wearer is. This kind of person's demand for self-actualization is also extreme in the performance of clothing. When physiological needs, safety needs, social needs, and respect needs are met, people will need to achieve self-actualization, manifested in the desire to maximize their abilities or potential through their efforts. For example, to achieve self-actualization through the pursuit of luxury brand clothing or private custom-made haute couture.

2.2.3 The relationship between Maslow's hierarchy of needs and clothing needs

By horizontal comparison, we found that the five symbolic attributes of clothing exactly match the five needs of Maslow's pyramid model (Table 2-1). The properties meaning of the two models from 1 to 5 levels are entirely reciprocal and synergistic (Figure 2-3). This discovery can lay a theoretical foundation for the follow-up research of interactive clothing subject. That is, the study of smart clothing should take the essential attributes of clothing as the theoretical support.

Table 2-1 Comparison of Maslow's hierarchy of needs and clothing needs

Level	Maslow's hierarchy of needs	Clothing needs
5	**Self-actualization**: realizing personal potential, seeking personal growth and peak experiences.	**Communication**: luxury fashion or haute couture, distinguish social status.
4	**Esteem**: achievement, mastery, independence, dominance, prestige, self-respect, and respect.	**Social connotation**: bring confident appearance; project thought, impressions and ambitions.
3	**Belongingness and love**: friendship, intimacy, affection.	**Image logos and tech development**: clothing displaying human personality and relationship with a group.
2	**Safety**: health and wellness, shading body, safety against accidents and injury.	**Modesty and protect**: consumers pay attention to the impact of clothing products on the body.
1	**Physiological**: for human survival, such as food, clothing, shelter and clothing.	**Warm and comfort**: clothing with a general function of shading and warming.

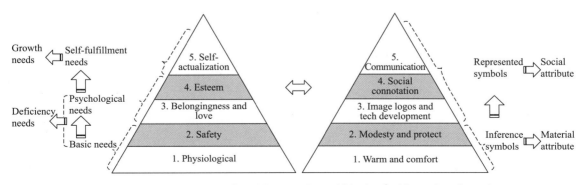

Figure 2-3 The synergy of clothing needs and Maslow's hierarchy of needs

2.3　The hierarchy and typology of clothing design

Bian（1998）, Waddell（2004）and Chevalier & Mazzalovo（2012）have explained why the clothing industry divides into three levels as Ready-to-Wear, High-grade Ready-to-Wear, and Haute Couture. As Waddell, G. once pointed out:

The levels of the fashion industry are actually the levels of manufacture, couture being the highest level and most time-consuming especially in terms of skilled labour, the number of repeat items being only in the tens or twenties. Ready-to-wear or pret-a-porter (in French) companies manufacture at a high level, hut industrially and in multiples of hundreds or thousands. Mass production is the cheapest and most highly industrially and can produce hundreds of thousands to millions of garments（Waddell, 2004, pp. X）.

However, slightly different from these scholars' views, this study integrated the design and production characteristics of clothing as well as the analysis perspective of the market distribution cluster hierarchy（Figure 2-4）, clothing can be divided into Ordinary Garment, Ready-to-Wear, High-grade Ready-to-Wear, High-grade Fashion and Haute Couture, five categories ranging from low level to high level as follows:

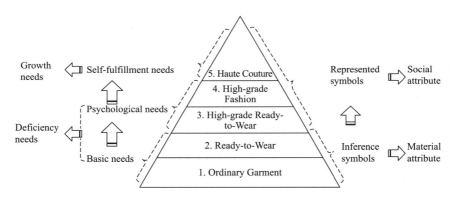

Figure 2-4　Hierarchy and typology of clothing design

（1）Ordinary Garment.

Ordinary clothing, which just provides the basic material function of clothing and does not attach to any design meaning, is also the cheapest clothing category for the mass market.

（2）Ready-to-Wear（Mass Ready-to-Wear）.

Ready-to-Wear products are clothes produced according to certain specifications and sizes. They are characterized by simplicity, beauty, practicality and short fashion cycle. It is generally divided into High-grade Ready-to-Wear and Mass Ready-to-Wear two types（Jiang, 2011）. Mass Ready-to-Wear clothes refer to the clothes produced in large quantities and standardized with the general public as the object of consumption. But there is also a relative lack of design characteristics.

(3) **High-grade Ready-to-Wear.**

High-grade Ready-to-Wear refer to clothes between high-grade fashion and Mass Ready-to-Wear, with the middle class as the object of consumption, small batches, and many varied styles. The difference between High-grade Ready-to-Wear and Mass Ready-to-Wear lies not only in their price, batch production and quality, but also in their design personality, taste and brand sense of belonging(Jiang, 2011).

Similar to Couture Fashion, High-grade Ready-to-Wear is also emphasis on custom made. The difference lies in that High-grade Ready-to-Wear emphasizes clothing can be produced in small batches of mechanized production and can be targeted at more consumers.

(4) **High-grade Fashion.**

High-grade Fashion, i.e., senior fashion customized by designer, while also emphasizing creativity and originality, embodying the characteristics of pop and fashion to provide customized design services for niche customers.

In the mid-20th century, with the rapid development of independent clothing industry in the United States, Britain, and other countries, Haute Couture also conforms to the trend of the times, and gradually derives High-grade Fashion clothing industry and High-grade Ready-to-Wear clothing industry.

(5) **Haute Couture.**

Haute Couture, the highest honor in the fashion industry, is only qualified to own a fashion workshop which has been awarded the title of "Haute Couture" by the Federation Francaise de la Couture. At present, there are only over 10 Official Members in the world such as Alexandre Vauthier, Alexis Mabille, Bouchra Jarrar, Chanel, Christian Dior, Christophe Josse, Franck Sorbier, Giambattista Valli, Givenchy, Jean Paul Gaultier, Maison Margiela, Stéphane Rolland, and Yiqing Yin.

The rules for Haute Couture membership are as follows: First of all, all fashion and accessories are designed and manufactured for private customers. Secondly, there must be a workshop in Paris, which has at least 15 full-time staff and employees more than three full-time models perennially and emphasize the use of manual completion of high-grade traditional crafts. Thirdly, the members should attend two high-end custom fashion exhibitions held by the Federation Francaise de la Couture in January and July each year, with no less than 35 sets of daily and evening clothes being released each season.

In France, Haute Couture enjoys a high social and cultural status and is considered the eighth art after literature, drama, painting, music, dance, sculpture and film. The Haute Couture has creative, artistic, ideological and forward-looking artistic characteristics.

Designers use the unique elements of clothing to create. They sublimate themselves with creative ideas, three-dimensional exaggerated styles, high-grade and novel fabrics, colorful layers, exquisite and complicated crafts, luxurious and beautiful decoration. The form and connotation of clothing, endow clothing with a new artistic perspective and expand the scope of modern art. It stands at the top of the fashion, its unique artistic tension and advanced aesthetics have a strong guiding role for the whole fashion trend.

The prices of Haute Couture are super high, and there are not many customers around the world who can afford Haute Couture, which proves that only Haute Couture can genuinely reach the highest level of "self- actualization needs" on clothing.

Similar to most industries, fashion requires innovation to survive and thrive; innovation depends on designers' creativity and original ideas to drive change, which is the reason for maintaining market continues to grow healthily and be attractive. As a result, Haute Couture, with its creativity, scarcity and high value-added, ranks at the top of the category hierarchy pyramid and leads the design of other types of clothing and market development.

These five levels are perfectly consistent with the clothing's symbolic attributes and Maslow's hierarchy of needs (Figure 2-5). This result verifies that the result of analyzing the hierarchy and category of clothing design is correct.

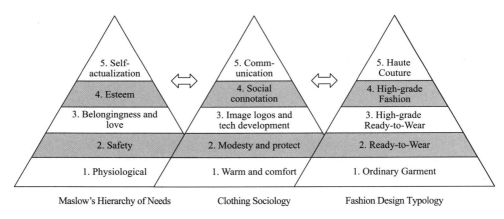

Figure 2-5　The synergy of Maslow's hierarchy of needs, clothing needs and clothing design

2.4　Deduce the evolution trend of interactive clothing—IoC

From an evolutionary perspective, interactive clothing has obvious potential applications. As can be seen from Figure 2-6, the future development of this field can be divided into three evolutionary stages according to my previous research (Wang et al., 2018a).

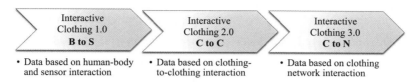

Figure 2-6 The interactive clothing evolutionary roadmap

- Interactive clothing in stage 1.0: Data on the human-body obtained via sensor monitoring (B to S). The clothing responds to the human physiology, such as the pulse, temperature, humidity, and even emotional reactions. At present, we can see some of the latest research results of healthcare clothing (Joo et al., 2016; Mana et al., 2016; Jagelka et al., 2016), these fruitful studies belong to the 1.0 stages of the evolution of smart clothing.
- Interactive clothing in stage 2.0: The data exchange based on clothing-to-clothing interaction (C to C). Using stage 1.0 as a basis, mutual reactions between two pieces or a series of clothing can be obtained to develop a new dynamic language of wearable expression, such as responses to the mutual distance and the other interactive reactions.
- Interactive clothing in stage 3.0: The data exchange based on clothing-to-network interaction (C to N). Based on stage 2.0, the interactions realized between multiple items or a series of clothing, thus allowing the formation of a clothing network. That is, stage 3.0 is the Internet of Clothes (IoC, reference to the concept of the Internet of Things) formed on the basis of the IoT technology platform. Through the 3.0 stage of interactive clothing, people can connect the virtual network and the real scene, carries on the free emotion communication, and thus constructs a smart clothing IoT as a carrier interactive chain to link wearer and things, things and things, wearer and wearer.

(1) **The association of IoT and IoC.**

IoT is a computing concept that contain electronics, software, actuators, and connectivity which allows these things to connect, interact and exchange data. It is describes the idea of everyday physical objects being connected to the Internet and being able to identify themselves to other devices. IoT involves extending Internet connectivity beyond standard devices, such as desktops and smartphones, to any range of traditionally dumb or non-internet-enabled physical devices and everyday objects. Embedded with technology, these devices can communicate and interact over the Internet, and they can be remotely monitored and controlled. When many objects act in unison, they are known as having "ambient intelligence." In other words, with the Internet of things, the physical world is becoming one big information system.

Source: Official website of techopedia

A thing, in the context of the IoT, is an entity or physical object that has a unique identifier, an embedded system and the ability to transfer data over a network. Given the prevalence of wireless technology, the increasing ability to miniaturize computer components and develop them inexpensively, that capacity could be developed for almost anything imaginable. In addition to dedicated computing devices such as PCs, smartphones or tablets, the list of potential things is almost unlimited. Clothing is of course also in this list, that is, internet of clothes (IoC).

(2) **The difference between IoT and IoC.**

Application innovation is the core of IoT development, user experience is the soul of IoT development.

- IoT is a technology interaction platform; IoC is an application innovation and technology extension of IoT, is a humanistic interaction platform.
- IoT is an information network that transmits technical information; IoC emphasizes the clothing network that transmits the symbols of humanistic emotion.
- IoT is mostly used in industrial production, while IoC emphasizes a future clothing lifestyle.
- IoT belongs to the technology system of technical category, and IoC is the humanistic evaluation system belonging to the social category.

Based on the above deduction, we can outline its development path from the perspective of clothing function as shown in Figure 2-7. It can infer the conclusion that the development path of clothing evolves from the ordinary everyday clothing to the functional clothing, the smart clothing, the interactive clothing and the internet of clothes. The focus of the combination of the smart clothing concept and IoT technology is in the display form of interactive clothing and IoC.

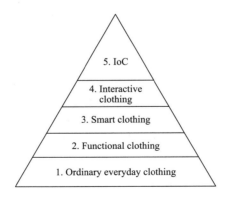

Figure 2-7　The evolution of clothing function

From the perspective of clothing function evolution (Figure 2-8), the pyramid level in this cyber world complete correspondence with the level of Maslow's hierarchy of needs, clothing sociology, psychology and market demand. All four categories evolved from low-level to advanced level, but within the horizontal comparison range, each level performs similar structure tasks and functions relationship. This verifies that the result of analyzing the evolution path of the clothing from a functional perspective is correct.

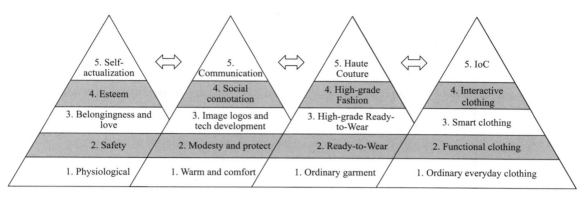

Figure 2-8 The synergy of clothing function categories with psychology, sociology and market deeds

2.5 The relationship between DIKWS and the evolution of clothing function

In the domain of information theory construction, when it comes to smart or interactive clothing, it is inevitable to trace to the IoT data theory hierarchy model, i.e., Data-Information-Knowledge-Wisdom-Service (Figure 2-9. DIKWS, for short) (Rowley, 2007; Frické, 2009).

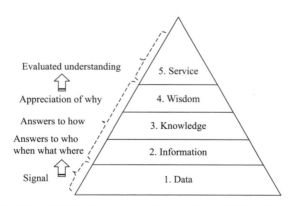

Figure 2-9 The DIKWS hierarchy (after Rowley, 2007; Huang, 2014)

Zins (2007) summarized the formulation of systematic conceptions of data, information, and knowledge is essential for the development of the information science and knowledge management.

Data is a series of disconnected facts and observations. These may be converted to information by analyzing, cross-referring, selecting, sorting, summarizing, or in some way organizing the data. Patterns of information, in turn, can be worked up into a coherent body of knowledge. Knowledge consists of an organized body of information, such information patterns forming the basis of the kinds of insights and judgments which we call wisdom. The above conceptualization may be made concrete by a physical analogy: consider spinning fleece into yarn, and then weaving yarn into cloth. The fleece can be considered analogous to data, the yarn to information and the cloth to knowledge. Cutting and sewing the cloth into a useful garment is analogous to creating insight and judgment.

Huang et al. (2014) expanded the previous research results and proposed the concepts of W2T cycle and wisdom as a service in conjunction with the cyber and Hyper world. The DIKWS refers to:

"The data transformation from raw data acquired in the physical world to information, knowledge, and wisdom in Cyber world, and the data circulation of Things → Data → Information → Knowledge → Wisdom → Service → Human → Things in the Hyperworld, combining the physical world and Cyber world... adopt a from data-to-wisdom data analysis and knowledge fusion platform called Wisdom as a Service to mine knowledge which leads to the wisdom for providing services from huge amounts of data. In this platform, there are three layers: structured data, information, and knowledge, and each layer can provide services as: data as a service, information as a service, knowledge as a service, and wisdom as a service on the whole."

Through horizontal comparison (Figure 2-10), we can find the corresponding relationship between DIKWS and the evolution of clothing function in each development stage from the level 1 to 5 (Wang et al., 2018b). The development process of data-information-knowledge-wisdom-service hierarchy and traditional-functional-smart-interactive-IoC (TFSII, for short) hierarchy is very consistent. They have the same attributes or complete the same types of the task at each development level (Table 2-2).

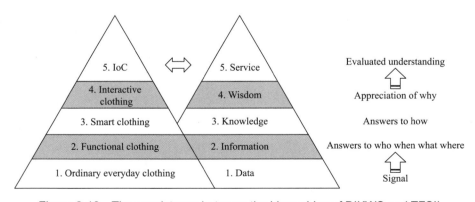

Figure 2-10 The consistency between the hierarchies of DIKWS and TFSII

Table 2-2 Compare the development process of DIKWS with TFSII

Level	DIKWS	TFSII
5	**Service**: combining the physical world and Cyber world	**IoC**: clothing-to-network interaction, network symbols
4	**Wisdom**: appreciation and evaluation of "why".	**Interactive clothing**: a carrier that forms the interpersonal interaction symbol in daily life.
3	**Knowledge**: application of data and information; answers "how" questions.	**Smart clothing**: embed CPS to form intelligent feedback.
2	**Information**: data that are processed to be useful; provides answers to "who", "what", "where", and "when" questions.	**Functional clothing**: wear in special environments to provide security, protection, and other functions.
1	**Data**: facts, individuals, signals, events.	**Ordinary everyday clothing**: style, fabric, color, signals.

2.6 The relationship between CPS/IoT and clothing

In the domain of information technology executives, when it comes to interactive clothing, it is inevitable to trace the architecture of CPS. Since Lee (2008) and Cardenas (2008) proposed the 3C infrastructure of CPS, the CPS architecture has been continuously improved. According to the architectural model proposed by Lee et al., (2014), Ning et al., (2016), Lee (2015) and Longo et al., (2017), the 5C architecture of CPS could provide conceptual guidance for the development of interactive clothing (Figure 2-11).

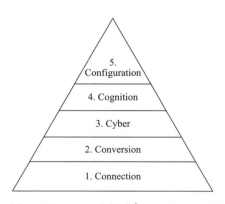

Figure 2-11 The 5C architecture of CPS (Lee et al., 2014; Ning, 2016)

Through horizontal comparison (Figure 2-12), we can find the perfect correspondence relationship between DIKWS, 5C architecture of CPS and the technical process of the interactive clothing at each development stage from the level 1 to 5, as details in Table 2-3.

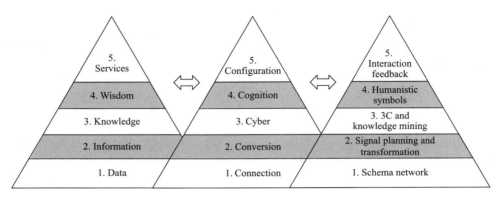

Figure 2-12 The consistency between the hierarchies of DIKWS, 5C and the technical process of the interactive clothing

5C architecture can provide conceptual guidance for interactive clothing development. At the same time, according to level 4 in Table 2-3, the humanistic design and evaluation of interactive clothing can inject humanistic vitality into the seemingly cold information technology, making it feasible for CPS to introduce people's daily emotional life. This part also responded to MRQ and SRQ2 mentioned in Chapter 1.

Table 2-3 Comparison of the technical process of the 5C and interactive clothing

Level	5C architecture of CPS	Technical implementation details of interactive clothing design
5	**Configuration**: self-configure, self-adjust, self-optimize; distributes the instructions.	Interpersonal interaction; symbolic-message; humanistic feedback to clothing.
4	**Cognition**: identify and make decisions, analyze task goals and status; collaborative optimization; knowledge generation.	Integrated simulation and synthesis of humanistic symbols; identify evaluation indicators; collaborative diagnostics and decision.
3	**Cyber**: data computing, distribution, decision and control of CPS system; data and knowledge mining.	Computing, communication, and control of clothing cyber performance diversity.
2	**Conversion**: Machine-based Algorithms; local intelligence; data conversion into useful information.	Signal planning, sensing, and transformation; specific algorithm.
1	**Connection**: sensor network; information collection and communication between devices.	Materials, style, and color; sensors, chips, and microprocessors.

From the signals sent by the wearer or clothing (Table 2-3, Level 1 and 2) to the humanistic symbols observed or perceived by the relevant parties (Table 2-3. Level 4 and 5), this evolutionary process is inseparable from the support of cyber. This is part responded to SRQ2 in Chapter 1. The evolution from information to knowledge requires the introduction of humanistic consciousness and symbols as a medium of evolution (Figure 2-12, Level 4). This also part responded to SRQ2 mentioned in Chapter 1.

This 5C architecture is still a conceptual model, not specific enough to guide the development of interactive clothing.

2.7 Other associated theoretical research

2.7.1 Technology-centered design and human-centered design

"Human-centered" originates from philosophical thought but not humanism. It refers to the most basic starting point of considering all the interests and needs of human beings in the design as the measure of the result of activities. Today's design community, the design of "human-centered" has almost become the "golden rule". The fundamental purpose of design is to deal with the relationship between human beings and things.

Krippendorff (2006) proposed several dimensions turn from technology-centered design to human-centered design:

"From the design of products to the design of artifacts that can play various social roles.

From a belief in technological progress to a concern for artifacts that are supportive of communities of users and are user-friendly for their individual members.

From culture-free conceptions of design in a universe to an acknowledgment of the role of language in the construction of diverse community-specific worlds.

From imposing intended functions of products to allowing people to use them in their own terms.

From designers as a lone genius or authority to designers who can work in teams to enroll the stakeholders of their designs in joint projects.

From attention to objects, products, material artifacts (an ontology), to an awareness of the processes of constructing and reconstructing (ontogenesis or design) artificial worlds whose sole purpose is to make sense to us, remain useful and enable us to feel at home with them".

Source: Krippendorff, K. (2006). The semantic turn: A new foundation for design. London, New York: Boca Raton. pp. 36

In the historical process of smart clothing development, the R&D concept which emphasizes technology as the core of scientific pursuit and technological intelligence effect has always been dominant. Krippendorff's point of view provides a clear guiding direction for solving the contradiction between technology and art design in the process of interactive clothing development.

2.7.2 Emotional design

Desmet (1999, 2009) and Norman (2004) have raised the research topic of emotion and design. Norman asserts that the emotional side of design may be more critical to a product's success than its practical elements. His emotional design model with three levels:

- Visceral (lowest level) is the level at which embodies the sensory aspects of how things look, feel, smell, and sound.
- Behavioral (middle level) is the level at which users form their perceptions of a particular product through use.
- Reflective (top level) is the level at which the product has meaning for consumers; it accounts for how consumers maintain an innate sense of identity through the consumption of the product over time.

Smart clothing also belongs to the field of artificial intelligence. Artificial intelligence can assist humans and help them perform tasks better, but still lacks a certain cognitive quotient and giving people a sense of apathy. The R&D of interactive clothing needs to inject humanistic emotion reflective elements.

2.7.3 Kansei Engineering

Kansei Engineering, founded by Professor Mitsuo Nagamachi, a methodology within the research field of affective engineering, aims to develop or improve products by transforming the customer's physical and psychological needs into the domain of product design parameters.

This method is helpful to understand consumers' emotions accurately, to reflect Kansei understanding in product design, and to create a system and organization for emotionally orientated design.

According to Nagasawa, one of the forerunners of Kansei Engineering, there are three focal points in the method: How to accurately understand consumer Kansei; How to reflect and translate Kansei understanding into product design; How to create a system and organization for Kansei orientated design.

Schütte et al. (2004) developed a general model covering the contents of Kansei Engineering, included the following steps: Choice of Domain, Span the Semantic Space and the Space of Properties synchronously, Synthesis, Test of Validity and Model Building.

There have been many successful cases in applying Kansei Engineering methods to clothing product development, so this study also adopts this method.

2.8 Summary

According to the above literature analysis and summary, it can be concluded that there are several hierarchies and typologies of clothing R&D from humanities and technology perspective.

Chapter 2
Literature Review, Definition and Deduction

2.8.1 Interactive clothing conforms to the evolution trend of clothing

On the development trend and evolutionary path of clothing (Figure 2-13), types of clothing property, grades of clothing property, and technical grades of functioning clothing property, they all support the future development of interactive clothing. According to the symbolic semantic model of clothing in Figure 2-11, clothing has multiple attributes or properties: ①Improving warmth and comfort; ②Providing modesty and protection; ③Displaying logos; ④Reflecting technological development; ⑤Conveying certain social and cultural connotations; ⑥Facilitating communication and expression.

　Ⅰ　Physical and Sociological properties of clothing: those six properties of the clothing mentioned above can be classified as three grades of physical (①, ② and ③), social (④ and ⑤) and spiritual properties ⑥ in Figure 2-13　Ⅰ.

　Ⅱ　Grades of clothing property according to Maslow's Hierarchy of Needs: Physical properties are also basic properties include (①, ② and ③). Sociological properties include (④, ⑤ and ⑥). Among them, ④ and ⑤ belong to intermediate properties, and ⑥ belongs to the advanced property (Figure 2-13　Ⅱ).

　Ⅲ　The technical properties are mainly combined with smart technology to achieve the technical functions of clothing. Each's needs should be satisfied at the lower levels before they progress to the higher, more complex levels. Technical grades of functional clothing property include three levels (see Figure 2-13　Ⅲ): ordinary everyday clothing belongs to basic grade, with (①, ② and ③) main properties and secondary properties (④, ⑤); smart clothing belongs to intermediate grade, with (④, ⑤) main properties and (①, ② and ③) secondary properties; interactive clothing belongs to advanced stage, main property is ⑥ and secondary properties are (④, ⑤).

Gepperth (2012) separated the application scenarios of wearable computing into three categories: "Sensing and data analysis", "Interfaces" and "Functionality and Aesthetics". In this research, I use the category of IoT functionality and design aesthetics to definite interactive clothing. Interactive clothing with the characteristics of fashion, one branch of smart clothes, is more emphasis on the characteristics of the clothing network with IoT system. Comprehensively, smart clothing mainly refers to technology upgrade based on the basic physical properties (①, ② and ③) of clothing. The interactive clothing mainly refers to enhancing the social properties based on smart technology (④, ⑤ and ⑥).

2.8.2　Humanistic and technical features of interactive clothing

Hierarchy and typology of clothing R&D from humanities and technology perspective. According to the horizontal comparison of humanistic and technical perspectives, interactive

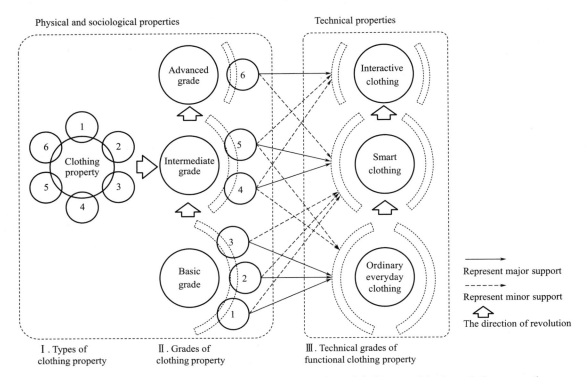

Figure 2-13　The interrelationships between properties of clothing and their evolutionary path

clothing is on the fourth level of two models respectively, and the transverse attribute of the level fourth is the function or characteristic of interactive clothing.

At the psychological level of humanistic perspective (Figure 2-14), interactive clothing needs to meet the "Self-esteem needs", and it is necessary to satisfy the function of "Connotation" in the sociology of clothing, and it belongs to the "Couture fashion" category in clothing design typology. This model part responses to the SRQ 1 mentioned in Chapter 1.

From the technical perspective (Figure 2-15), interactive clothing needs to meet "Wisdom" requirements in IoT data theory, "Cognition" functionality in the 5C architecture of CPS, and the transmission of "Humanistic symbols" regarding technical processes. This model also partial responses to the SRQ 1 mentioned in Chapter 1.

The relationship between each level of the different category in Figure 2-15 as follows:

All four categories evolved from low-level to advanced level, but within the horizontal comparison range, each level performs similar structure tasks and functions relationship.

In the context of the level 1, they are representing stimuli or signals, objective facts or observations, which are unorganized and unprocessed and therefore have no meaning or value because of lack of context and interpretation. They are "of no use until…in a usable (that is, relevant) form" (Zins, 2007; Rowley & Hartley, 2008).

In the context of the level 2, they are inferred from level 1 and differentiated from level 1 in that they are "useful" in the process of answering interrogative questions (e. g., "who" "what" "where" "how many" "when")(Bellinger, Castro, & Mills, 2004), thereby making the data useful for "decisions and action" (Liew, 2007). Level 2 is defined as they are endowed with meaning and purpose.

In the context of the level 3, they are defined with reference to level 2. Definitions may refer to the context of level 2 having been processed, organized or structured in some way, or else as being applied or put into action.

In the context of the level 4, they are described as "know-why" (Zeleny (2005), so as to differentiate "why do" from "why is", and expanding the definition to include a form of know-what ("what to do, act or carry out"). And these will bring enlightenment to the level 5, as well as represents evaluated understanding.

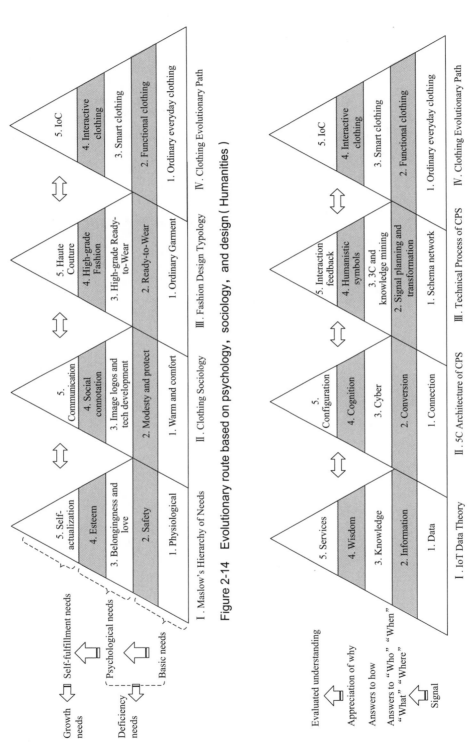

Figure 2-14　Evolutionary route based on psychology, sociology, and design (Humanities)

Figure 2-15　Evolutionary route based on IoT and CPS applications (Technology)

Chapter 3
Research Model and Methodologies

3.1 Research roadmap

According to the research aims, objectives, and the research questions of this study mentioned in Chapter 1, if we want to explore what happened in the process of realizing interactivity of interactive clothing and what technical means were used in this process, then we need to disassemble and analyze the technological flow of interactive clothing R&D.

The research roadmap, according to Soft System Methodology, i.e., the "WHY" "HOW" and "WHAT" shown in Figure 3-1. The factor "WHY" investigates the reasons why interactive clothing should satisfy technical functions and human emotional expression simultaneously. This problem cannot be separated from the grades of clothing symbolic properties mentioned in Figure 2-11, that is, clothing has multiple attributes or properties: ①Improving warmth and comfort; ②Providing modesty and protection; ③Displaying logos; ④Reflecting technological development; ⑤Conveying certain social and cultural connotations; ⑥Facilitating communication and expression.

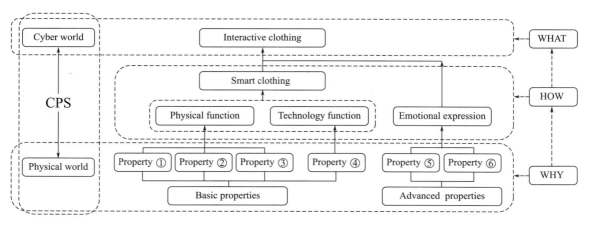

Figure 3-1 Research roadmap

"HOW": This factor refers to disassemble and analyze the technical process of interactive clothing design and development, investigates the manner in which artistic design perspectives and engineering methods can be effectively combined, blending emotional communication with IoT/CPS technology in the form of interactive fashion innovation.

"WHAT": This factor refers to the R&D elements of interactive clothing in the context of cyber world based on IoT/CPS technology. Therefore, the research roadmap should revolve around the CPS' deep interaction between the physical world and the cyber world, and explore the relationship between why, how and what factors.

3.2 Creating the CPCS research model

Using the three factors "WHY" "HOW" and "WHAT" mentioned above, an innovative system architecture is developed for interactive clothing.

The technical process of interactive clothing design and development can be summarized from the three models (Figure 2-13-Figure 2-15) mentioned in Chapter 2. The technical relevance of interactive clothing, DIKWS, and CPS was revealed in the Cyber-Physical-Clothing Systems (CPCS) model (Figure 3-2). Response to the MRQ of this research, this model can explain what happened and how do the social semantic of the interactive clothing transform from the information stage to the knowledge stage in the architecture of CPS.

The role of this model is to explain the technical development process of interactive clothing from clothing to data, information, knowledge, wisdom, services, humans, and then back to clothing (the Clothing to Human cycle, C2H for short) in the perspective of CPS and DIKWS architecture.

The interactive clothing has the characteristics of information processing and human feedback, it is controlled by machine-based algorithms and integrated the network with its wearers into one unified and holistic framework, which determines its interactive architecture, like other scholars' research results, can be divided into three levels: physical level, cyber level and social level (Sheth et al., 2013; Rajkumar et al., 2010). The vertical direction in this model represents the functional relationship of progressive interaction from physical level to cyber and social level, and the horizontal direction corresponding is the technical principle relationship in three levels between C2H technical implementation (inside the blue dotted box), DIKWS, and CPS loops (inside the red dotted box). The dotted double arrows represent interactions and effects, and solid line wide arrows represent development directions.

3.2.1 Physical level: entity item of clothing

(1) This level is implemented by embedding the sensors and microprocessors in combination with the style design c to create a network, performing the data connection with the clothing material or devises and forms particular signals, and finally implementing the signal conversion to the cyber level utilizing sensing.

(2) The characteristic of the process from step ① to ②③ is the information process from the fundamental data on the physical level of clothing to the cyber level.

Chapter 3
Research Model and Methodologies

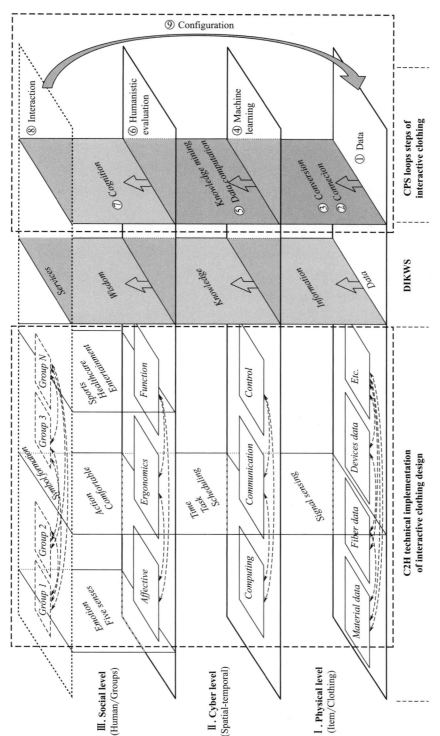

Figure 3-2 A CPCS architecture for interactive clothing in the hyper world

(3) The contents of this process.
  a) Among them, the process from step ① to ② belongs to Intellisense. The core of this process is the creation of a CPS architecture that allows data to be measured and aggregated in various forms, and the communication between embedded sensors and other devices is also possible.
  b) Among them, the process from step ② to ③ is information mining. The core of this process is the Machine-based Algorithms on the device side so that some data can be analyzed and utilized locally in the device to achieve local intelligence.
(4) There are also interactions between different categories of data.
(5) The technical difficulty at this level is the transformation from connection to conversion.

3.2.2　Cyber level: spatial-temporal concept of clothing

The cyber level represents the spatial-temporal concept of clothing in the process of interactive information processing. It is the information technology processing stage in the process of transforming physical signals into humanistic symbols.

(1) The cyber level, which covers steps ④ and ⑤ in the CPS process, is the core of data processing, distribution, decision making, and scheduling control of the entire clothing CPS system. Here, a big data environment originated from clothing or wearer, while running advanced analysis algorithms for large-scale computing and knowledge mining.

(2) The content of this level is the organic integration and deep cooperation of Computing-Communication-Control employing time, task and scheduling, and achieves a deep combination of the cyber and physical world involving the object mechanism, environment, and group.

(3) The characteristic of this level is the process of transforming the signaling of clothing data into the symbolization of humanities knowledge.

(4) The technical difficulty at this level is the knowledge mining based on data computation.

3.2.3　Social level: humanities activities with interactive clothing as the medium

Social level represents the humanities activities, interpersonal interaction and communication between individuals or groups (Ning et al., 2016) with interactive clothing as the medium.

(1) This level includes steps ⑥, ⑦ and ⑧, the main role is to evaluate, identify and feedback, decision-making for the interactive activities of clothes. By analyzing the task objectives and status exhibited by each piece of clothing in the current system, an interactive decision-making of collaborative optimization among multiple pieces of clothing are established.

(2) Its main content is humanistic evaluation and cognitive formation. At present, the design evaluation indicators for smart clothing or interactive clothing are mainly divided into three dimensions, that is effective indicators (emotion, five senses, etc.) (Chen et al., 2017),

ergonomics indicators (action, posture, comfortable, etc.) (Hsiao et al., 2015; Kim et al., 2010), and functional indicators (sport monitoring, healthcare, entertainment, etc.) (Chen et al., 2016; Kim et al., 2016a; Salim et al., 2014). The follow-up prototype evaluation in this study was adopted by the emotional/Kansei evaluation method.

(3) The technical difficulties are the formation of cognition after performing group differentiation evaluation and the guidance of symbolic interactive feedback. Because of the interaction among different wearer groups, it is necessary to meet the differentiated needs of different types of dress groups, as well as the differences in various issues and attributes among each wearer group. Different types of feedback forms are represented by different clothing external symbols. Various communication symbols that can express social message to be output during the interaction of different groups or individuals.

(4) This process is characterized by the transformation of clothing from knowledge to wisdom and social services, and clothing becomes the carrier of automatic feedback of the wearer to the associated person or clothing.

3.2.4　Cross-level: transform from cyber to social structure

The configuration execution layer ⑨ runs through the three levels of physical, cyber and social. After receiving the decision, the configuration execution layer transforms the decision into instructions that are understood by the logic of each subsystem, and finally performs some external performance reaction of clothing by the microprocessor and sensor embedded in the clothing.

3.3　The definition of research methods

3.3.1　Practice validation and prototyping methods

(1) Through the literature induction and deductive research methods mentioned above, the CPCS model was created.

(2) First of all, it should practice verifying the feasibility and rationality of the application of IoT technology in clothing.

(3) Practical verification of the possibility of clothing interaction, and ascertains the viability of the CPCS model.

(4) Practice verifying the diversity of clothing interaction, from personal interaction to the scale of group interaction.

3.3.2　Prototype evaluation method

(1) Identification of authoritative institutions, that is, objective evaluation method.

(2) Kansei Engineering evaluation method, that is, subjective evaluation method.

3.3.3　Model inductive method

（1）Analyze the design process and results of prototypes, and construct the design principles of interactive clothing.

（2）Analyze the process and results of the evaluation, create interactive clothing evaluation criteria system, and propose qualitative and quantitative evaluation indicators.

In summary, this study will mainly adopt a variety of research methods from Deduction to Induction, Empirical, Evaluation and Abduction in a logical order. The logical relationships of research methods and verification forms are shown in Figure 3-3. The CPCS model is used to guide the prototype development, and the prototype is evaluated subjectively and objectively, and a new guiding model is built based on the evaluation results and the effect of the prototype. Prototypes are used to validate the CPCS model, the purpose of the evaluation is to guide the improvement of prototypes, design principles are used to validate prototypes, and evaluation systems, in turn, validate evaluation methods.

3.4　Summary

According to the research aims, objectives, and the research questions of this study, the research roadmap of "WHY" "HOW" and "WHAT" is proposed by Soft System Methodology.

An innovative CPCS systems architecture is developed for interactive clothing. This model is the most critical research idea for prototype development practice in this study, which lays a theoretical foundation and practical guidance for the follow-up practice research.

The technical process of interactive clothing design and development was summarized. The technical relevance of interactive clothing, DIKWS, and CPS was revealed in the CPCS model. The role of this model is to explain the technical development process of interactive clothing from clothing to data, information, knowledge, wisdom, services, humans, and then back to clothing (the Clothing to Human cycle, C2H for short) in the perspective of CPS and DIKWS architecture.

The research methods mainly adapt deduction, induction, empirical, evaluation and abduction, and also include prototyping method, Kansei Engineering evaluation method, model verification method, and other methods of quantitative research and qualitative research. The prototype will be developed following the CPCS architecture and verified by the Kansei evaluation method, and the design principle and evaluation system of the interactive clothing will be deduced at last.

Chapter 3
Research Model and Methodologies

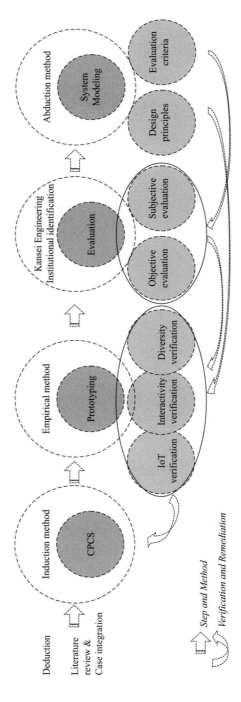

Figure 3-3 The logical relationships of each research method

Chapter 4
Prototyping of Interactive Clothing

The practical research of this topic was started in 2012, and progressive results were getting during three stages. The first stage was in 2013, two types prototype of infant's smart clothing was developed. The second stage is which interactive couple clothing has been made from 2016 to 2017. The development of interactive parent-child clothing is the third stage from 2017.

The reasons for choosing these three cases as empirical examples of this study are:
- Case study 1, smart infant clothing. Analyze the feasibility of the application of IoT technology in the design and development of garment products from the perspective of sensor technology application.
- Case study 2, interactive couple clothing. It is a preliminary study for IoC. To bridge the gap between human emotions and wearable technologies for interactive fashion innovation, to consider the reasons why smart clothing should satisfy the IoT technical functions and human emotional expression simultaneously, to investigate the manner in which artistic design perspectives and engineering methods combined effectively.
- Case study 3, interactive parent-child clothing. To bridge the gap between CPCS architecture model and emotional evaluation, to explain how the transformation could be realized from the information to common feeling, to refine the design elements for interactive clothing.

4.1 Case study 1, smart infant clothing

To verify the feasibility of IoT sensing technology applied to clothing, this research completed two types of infant's smart clothing development in 2013. As a result, two Chinese utility model patents have been granted, which show the feasibility and rationality of this case study. Furthermore, the fundamental study of interactive clothing has established the technical foundation by this experiment.

4.1.1 Infant's crotch humidity monitoring alarm trousers

The purpose of the experiment is to combine IoT sensing technology and infants clothing design, to develop a real-time humidity monitoring alarm the crotch trousers for infants. According to the supernormal change of the crotch humidity, the children's guardian will receive an alarm.

4.1.1.1 The technical solutions

Referring to Figure 4-1, there are humidity sensor 2, signal wire 3, chip system 5, flexible zipper 4 and trousers crotch parts. Among them, the humidity sensor 2 is fixed on the crotch of infant trousers 7. Chip system 5 is put at the foot of infant trousers 7. Humidity sensor 2 is connected with chip system 5 by signal wire 3. Flexible zipper 4 is set at the crotch of infant trousers 7. Humidity sensor 2, signal wire 3 and chip system 5 are sewn into the hollow seam of inside position of infant trousers 7. They can be installed or taken out freely. The crotch spare parts are placed in the trousers bag 1 at the rear position of trousers.

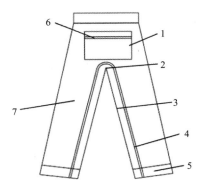

Figure 4-1　The structure of trousers and electronic components

The chip system 5 is equipped with a wireless sensor and wireless connection with the receiving device. The receiving device is equipped with a humidity test and displays an alarm.

When the infant's trousers are using, it is necessary to adjust the position of the humidity sensor 2 and pull the flexible zipper 4. In the event of the humidity exceeds the standard, an alarm sound will be issued, at the same time, the humidity value will be displayed digitally to remind timely replacement. The replacement process is simple, firstly the children's guardian should open the flexible zipper 4 to remove the wet diaper, then open the zipper 6 of the pants pocket, and take out the diaper from the trouser pocket 1 for replacement.

4.1.1.2 Value and significance

The humidity sensor is used to monitor the humidity of the crotch of infant's trousers at all time. As soon as the humidity was out of the setting value range, an alarm will be given by the receiving device. Meanwhile, the infant's trousers need not be replaced entirely. It only needs to open the flexible zipper to replace the wet diaper. The whole device is set in trousers and can transmit the humidity status data to the guardian through wireless signals. It avoids many tedious things in the process of monitoring infants and provides a better environment for the growth of infants, to facilitate the healthy growth of infants. The prototype and the patent certificate are shown in Figure 4-2.

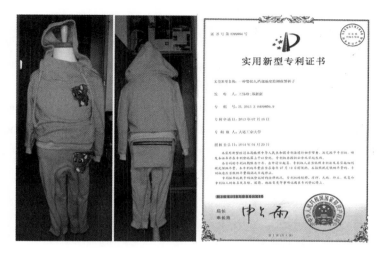

Figure 4-2　The prototype and the patent certificate 1

4.1.2　Temperature monitoring infant clothing

This purpose of this experiment is to design a real-time temperature monitoring system based on IoT sensor technology and to remind guardians to take appropriate measures according to the temperature status of infants, to better nurture infants.

4.1.2.1　The technical solutions

Referring to Figure 4-3, there are heat-sensitive sensor 3, chip system 1, signal wire 2, underarm garment structure 4 and hollow garment seam 6. Among them, heat sensitive sensor 3 is placed in an underarm garment structure 4 of the jacket 7. Chip system 1 is installed in the hem position of the jacket 7. Signal wire 2 is set between heat sensitive sensor 3 and chip system 1. Signal wire 2 is placed in hollow garment seam 6. There is a wireless transmitter on chip system 1. The structure of electronic components can refer to Figure 4-1.

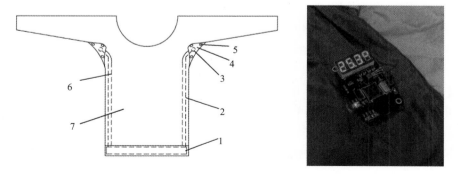

Figure 4-3　The garment structure and electronic components

There are two positions of the heat-sensitive sensor 3, one is arranged in the left and right underarm garment structure 4, and the other is connected with the chip system 1 by signal wire 2. A button 5 is arranged on the underarm garment structure 4 as shown in Figure 31, which can be disassembled and installed for easy replacement and cleaning. The hollow garment seam 6 is made of the front and back garment pieces. The receiving device is placed beside the guardian, which is a sound alarm and a temperature display device that can receive wireless transmission signals.

In view of the defect of temperature detection accuracy and lack of sensitivity in the design method of non-contact body temperature detection embedded in the armpit of clothing, 10 male and 10 female children were invited from two kindergartens to carry out temperature test experiments, the body temperature of the armpit of 20 children (from A to T in Table 4-1) was extracted in four states of quietness, exercising, after exercising and sleeping. Finally, the mean temperature 37.25℃−2.25℃= upper limit 35℃ and lower limit 29℃ was taken as the interval warning response threshold of the prototype.

Table 4-1   Temperature test extraction record

Status	A	B	C	D	E	F	G	H	I	J
Quietness	36.8℃	36.5℃	36.9℃	37.1℃	37.0℃	36.9℃	36.8℃	36.9℃	36.9℃	37.1℃
Exercising	37.2℃	37.1℃	37.3℃	37.2℃	37.4℃	37.5℃	37.2℃	37.4℃	37.3℃	37.6℃
After Exercise	37.7℃	37.4℃	37.8℃	37.9℃	37.9℃	38.1℃	37.7℃	37.9℃	37.8℃	38.2℃
Sleeping	37.1℃	36.7℃	37.2℃	37.3℃	37.2℃	37.0℃	37.1℃	37.2℃	37.0℃	37.3℃
Status	K	L	M	N	O	P	Q	R	S	T
Quietness	37.0℃	36.7℃	36.9℃	36.7℃	36.7℃	36.8℃	37.2℃	36.9℃	36.9℃	36.6℃
Exercising	37.2℃	37.0℃	37.1℃	37.1℃	36.9℃	37.1℃	37.4℃	37.3℃	37.5℃	36.9℃
After Exercising	37.8℃	37.7℃	37.6℃	37.9℃	37.6℃	37.8℃	37.9℃	37.8℃	38.1℃	37.5℃
Sleeping	37.2℃	36.9℃	37.3℃	36.9℃	37.0℃	37.1℃	37.3℃	37.2℃	37.4℃	36.8℃

4.1.2.2 Value and significance

The caregivers arrange the position of heat-sensitive sensor 3 as shown in Figure 30 which is under the armpit of infants. Then caregivers should open the receiving device and place it in the closed position of caregivers. When the temperature exceeds a set value range, the receiving device will give an alarm to remind the caregiver to do further processing. The prototype and the patent certificate are shown in Figure 4-4.

4.2 Case study 2, interactive couple clothing

This case study was implemented from 2016 to 2017, and the project research carrier is interactive couple clothing.

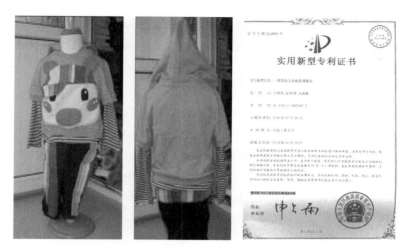

Figure 4-4   The prototype and the patent certificate 2

4.2.1 Determining the categories of experiential prototype

Product design should combine with consideration of the future market demands. The future main target consumers of smart clothing are expected to be college students and other young people who are willing to accept new fashion things. In 2016, we conducted a questionnaire survey distributed to 369 students of two universities in Dalian, China (Weizhen et al., 2017), participants were studying one of five disciplines (①artistic design, ②materials science, ③literature, ④ management, ⑤information science), the demand rate for interactive clothing among couples was found to be more than 55%, which is significantly higher than the demand for other types of clothing (Figure 4-5).

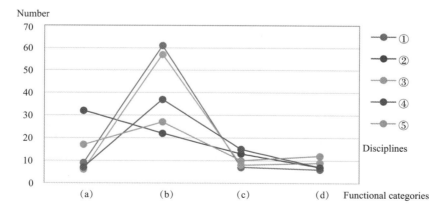

Figure 4-5   Survey data for prototype selection

Note that the survey participants were asked to select from the main functional categories of smart clothing: (a) medical-care clothing; (b) interactive clothing for couples; (c) sports monitoring clothing; (d) nocturnal road-safety clothing. Based on the findings, option (b), i. e., interactive clothing, was selected as the categories of experimental prototype.

4.2.2　Purpose, methodology and findings

- The aim of this case study is to bridge the gap between human emotions and wearable technologies for interactive fashion innovation. To consider the reasons why smart clothing should satisfy the IoT technical functions and human emotional expression simultaneously, to investigate the manner in which artistic design perspectives and engineering methods combined effectively, to explore the R&D elements of future smart clothing based on IoT technology.

- This case study combines artistic design perspectives with information-sensing engineering methods as well as Kansei evaluation method. Micro-sensors and Light Emitting Diodes (LEDs) embedded in couples clothing prototype. The first experiment steps in the design and production of prototype clothing, and do the initial emotional evaluation. The second experiment is a comparative evaluation of the prototype and other typical smart clothing.

- The interactive clothing prototype was shown to correlate well with human emotional expressive patterns. The evaluation I indicated the prototype could stimulate the emotional response of the participants to achieve a higher score in the activate sensor state. Evaluation II revealed that in the process of interactive clothing design, the technical functionality should synchronize with the requirements of human emotional expression.

4.2.3　Hypotheses and methodology

4.2.3.1　Hypotheses

With the unremitting efforts of researchers, we have preliminarily realized the function of making the clothing smart. However, in the realization of technological progress, it is necessary to explore the possibility of integrating smart clothing into the daily life of consumers and the future development direction.

- *Hypothesis 1. Interactions may occur between two or more items of smart clothing.*

  At present, the research of smart clothing mainly focuses on the use of clothing to reflect physiological data from the human body, and the majority of individual items of clothing respond to a single person. However, the development of multiple pieces of clothing worn by different individuals and having a mutual reaction relationship should also be possible.

- *Hypothesis 2. Interactions between items of smart clothing can well match with the emotional relationship of the wearers.*

  If multiple pieces of clothing have a reciprocal relationship, the primary emotional responses of the body can be reflected through this relationship. Smart clothing can reflect

the relationship such as happiness, anger, sadness, and joy between wearers.
- *Hypothesis 3. As a new type of clothing, emotional interactive clothing can represent one of the main development directions in future daily smart clothing lifestyle.*

The future smart clothing lifestyle that people expect was predicted to occur in two distinct directions: performance-driven and fashion-driven (Lam Po Tang and Stylios, 2006). In the field of fashion design, aesthetic and emotional design expressions are detached from the physical, social and spiritual attributes of clothing. The artistic aesthetics and human emotion characteristics that conform to the desired lifestyle of the target consumer group. Emotional interaction and transmission between people or between people and artifact may be more critical to smart clothing's success than its practical function elements.

4.2.3.2 Design methodology

The methodology employed in this study combines clothing style design, i.e., artistic design perspectives, with information-sensing engineering, i.e., engineering methods. This corresponds to the second design factor "How" introduced above. Micro-sensor elements embedded in clothing and effective integration obtained, such that the clothing worn by two couples exhibits a mutual relationship, as expressed through a connected performance using LEDs or color effects. The reason for choosing LEDs and their color variations into design elements is that the visual effect of the LEDs is utilized. Because the three significant elements of clothing product design, that is, the style, color and material belonging to the visual category, and the LEDs illumination effect is noticeable.

4.2.4 Experimental approach

The experiment divided into two steps. The first step in the design and production of prototype clothing, and do the initial emotional evaluation. The second step is a comparative evaluation of the prototype and other typical smart clothing.

4.2.4.1 Prototype style and design basis

The prototype is interactive cold clothing for couples. The clothing items contain cold-proof liners with cotton filling, which can freely replaced by using differently colored materials, suitable for a resource-limited society. Thus, one item of clothing can exhibit various colors, embodying the concept of energy-intensive design and providing a demonstration for realizing 5R (Reduce, Reevaluate, Reuse, Recycle, Rescue) design goal (Wang et al., 2014). The prototype design styles can be seen in Figure 4-6 (labeled a-d from left to right for future reference). Among them, "a" and "b" are two women's styles, "c" and "d" is a couple of styles.

According to the conclusion of brain science (Pan et al., 2017; Marazziti et al., 2017), the romantic relationship between lovers is a typical representative of human emotions, so we selected lover couples (not for random people or co-workers) as prototype target user of interactive clothing design.

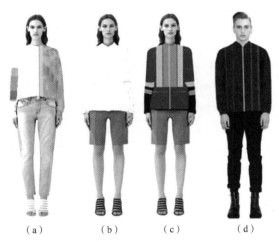

Figure 4-6　Clothing designs

From the psychological point of view (King-O'Riain, 2016), the distance can also represent the intimate or estranged relationship between two persons. Therefore, we decided to use range as a trigger for the interaction, and embed the sensor into the clothing. To hide the sensing elements in clothing, we selected the relatively heavy winter clothes as a prototype design program.

In the external display design of the sensing reaction, the prototype is intended to draw on the more mature LEDs display technology (Chun and Lee, 2016; Rapisarda, 2016; Leonard, 2017).

4.2.4.2　Material selection

• **Clothing outer fabric and liner filler material**

Transparent material is the material of choice for research on technology in fashion design.Therefore, the outer fabric of the prototype developed in this study was thermoplastic polyurethane. The liner filler material was a product known as "Wool Felt Poke Fun" in China, which is widely available in the fabric market and could be freely replaced using different material colors.

• **Sensor material**

The sensors were comprised of WH-335-DT and 5050-60RGB (Hongjing tech Co., Ltd., Shenzhen) LEDs in the form of a flexible flat ribbon that could be bent into any shape. Note that, by using a special clip, these ribbons can be fixed in the shapes of various letters or patterns to meet different design requirements. The LEDs ribbons are quite small that a unit can be cut to contain only three diodes. Thus, these LEDs ribbons can easily be installed in very narrow parts of clothing. In this study, 5-8 LEDs embedded in every item of clothing.

A US100 (Telesky Electronic Co., Ltd., Shenzhen) ultrasonic sensor module was also used, which has a 5-m non-contact range function for both general purpose input/output (GPIO) and serial communication. Considering the compatibility of traditional code programs and the operating speed, we also employed a STC12C5A60S2 (Hongjing tech Co., Ltd., Shenzhen) single-chip microcontroller. This microcontroller has ultra-high speed and low power consumption and is fully compatible with the traditional 8501 instruction code. A mic-transformer was also required, to regulate the operating voltage between the different modules. Note that the above electronic components are widely available in the electronics market, therefore, this design is suitable for a resource-limited society.

- **Electronic embedding design**

Take style "a" as example, the battery ① and single-chip ② are hidden in the inside pocket, and sensors ③ embedded in the back center seam of the clothing, using insulated wire ④ hidden in the hem and side seams (Figure 4-7). Schematic circuit design which takes 5 LEDs as the example can be seen in Figure 4-8. The US100 serves as the main creative medium, and uses a single-chip programming process to control the LEDs brightness changes. The entire circuit access voltage is 12V, US100 operating voltage DC 2.4V - 5.5V, STC12C5A60S2 operating voltage is 3.3V, so the circuit also needs to add a transformer. The above materials and the Kansei evaluation described below correspond to the "WHAT" design factor presented above.

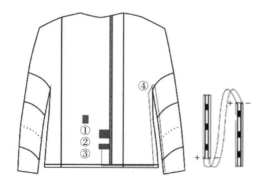

Figure 4-7　Electronic embedding design for style "a" in Figure 4-6

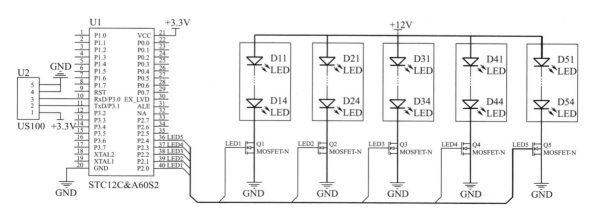

Figure 4-8　Schematic circuit design

4.2.5 Results

By realizing the interaction between two clothing items, the interactive clothing was shown to correlate well with human emotional expression patterns (Figure 4-9). When the distance reduced between two individuals wearing the prototypes, the LEDs embedded in the two clothing items gradually illuminated. The illumination became apparent at a separation of 4.5 meters, with full brightness been obtained when the wearers were separated by less than 1 meter.

In Figure 4-9, style "a-1" is the effect of style "a" activated sensor, the difference between "a-1" and "a" is the LED effect in the right chest and left sleeve (shown in the red dotted circle). Style "b-1" is the effect of b-activated sensor; "b-2" is the effect of the dark environment (shown in the red dotted circle). Style "c" and "d" two items clothing together are couple style, and "d-1" is a close-up of the vertical bar-shaped LED effect in the front chest position.

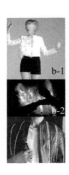

Figure 4-9　Photographs of prototypes visual effects
Notes: Examples of style a (left), style b (center), style c and d (right)

4.3　Case study 3, interactive parent-child clothing

4.3.1　Purpose, methodology and findings

This case study was implemented from 2017 to 2018, and the project research carrier is interactive parent-child clothing.

- The aim of this case study is to bridge the gap from a transdisciplinary perspective between CPS architecture in the field of information science and emotional evaluation in the field of humanistic science for interactive fashion innovation, to explain how the transformation could be realized from information to common feeling in the process of interactive clothing design.
- By using the CPCS architecture model in the hyper world, and taking the development process of interactive parent-child clothing as a case to analyze the transformation from the physical signal input to the social symbol recognition output, this study verifies the feasibility of the CPCS model adopted in interactive clothing design.

- This case study provides a fruitful practical application reference for designers who are engaged in the art design field but not familiar with relevant information technology, and explains the application principle and technical process of CPCS for interactive clothing development.

4.3.2 Background and research question

The focus of Human Computer Interface (HCI) has moved looks to the future of technology and how it can enhance human activity and experience. As a multi-dimensional smart mechanism that tightly integrated computing, communication and control (3C) with a deep interaction between physical and cyber world, the CPS will provide HCI with the foundation of emerging and future smart services, and improve our quality of life in many areas (Duarte Filho et al., 2015; Huang et al., 2018; Barnaghi et al., 2015). Following the ascend anting application of CPS to the fashion industry, the academic and industry research concerns in this area across various disciplines stems have mainly been devoted to optimizing the interaction of smart clothing by the integration of information technology. Therefore, as a development branch of smart clothing and also one symbolic medium in the field of sociology (Kaiser, 1996), interactive clothing emphasizes the interactive characteristics of communication and expression between human and objects (Karrer et al., 2011; Wang et al., 2018b), and is adopted as the target carrier of this research project.

We are still confronted with several challenges from correlative information technology and humanities technology respectively.

Firstly, when we extend the functional definition of interactive clothing to provide interaction between different wearers or groups rather than just personal interaction with data, it is an important challenge to apply CPS in the development of interactive clothing. That is, how to meet the differentiated needs of different wearers and the differences in various attributes among wear groups. In addition, most of the existing CPS architecture research is based on the logic framework of information science, but not the specific technical guidance for the non-information engineering personnel. Therefore, it is very important to optimize the CPS technical architecture which should with wide applicability for interactive clothing (Chen et al., 2016a; Alhafidh et al., 2017; Longo et al., 2017).

Secondly, current researchers seldom dabble in the systematic research of smart clothing from information modelling to humanistic evaluation, most of them analysed the information technology application or humanistic design from the single academic background (Trindade et al., 2016; Kim et al., 2016a; Salim et al., 2014; Weizhen et al., 2017), which may lead their inability to systematically explain the principle and process of how to extract data, transform and form humanistic feeling of the interactive clothing.

The aim of this case study is to bridge the gap between CPS in the field of information science and emotional evaluation in the field of humanistic science, to systematically explain how the transformation could be realized from the information on knowledge in the process of interactive clothing design. By optimizing the CPS architecture of interactive clothing and conducting the Kansei Engineering (Nagamachi, 2011b) analysis of wearers' evaluation, to explore the R&D elements and suggest practical implication for interactive clothing design.

To attain the above objectives, this case study will answer one Major Research Question: What happened and how happened when the interactive clothing in the interactive transformation process from the information to knowledge stage? In addition, the Subsidiary Research Questions (SRQ) as follows. SRQ1: How is the connection could be realized between the signals in the field of information physics and the symbols in the various humanities and social dimensions? SRQ2: How is the humanistic evaluation could be realized the optimization in the stage of knowledge formation?

4.3.3 Prototype design following the CPCS model

Through the deductive analysis of the models in Figure 2-14, Figure 2-15, and Figure 3-1, we should realize that the R&D trend of the interactive clothing would move from the function of one piece of clothing to the interaction of multiple clothes approaching the CPS era. Next, the interactive clothing prototype is designed and demonstrated in accordance with the CPCS model.

As a symbolic concept clothing, the parent-child clothing is one perfect symbolic carrier of the emotional design of clothing, which can create a micro-hyper world to express the integration of family belonging and clothing culture, emphasizes parent-child interaction in everyday family life and family circle's emotional relations (Keel, 2016), sublimated into a spiritual category and cultural awareness (Brighouse & Swift, 2014). The interactive clothing conforms to the CPCS' technical conditions in which the interpersonal signal output interacts with the social symbol feedback of the family atmosphere. Therefore, we have reason to infer that the introduction of CPS into the design and development of parent-child clothing, and employ the concept of local area network to show the family atmosphere and spatial-temporal concept, which will be an exciting and potentially promising research topic, also provides a practical approach to realize the interactive clothing for the coming age of ubiquitous intelligent CPS applications.

4.3.3.1 The prototype development approach

Regarding style design, it takes the form of a three-piece family clothing combination of parents and children.

Concerning technical application, the sensing technology is used to create a family LAN system for three-piece clothing.

Regarding the appearance of the garment, the visual effect of the LED is utilized. Because the three significant elements of clothing product design, that is, the style, color and material belonging to the visual category, and the LED illumination effect is noticeable.

4.3.3.2 Design methods of interactivity

The first one is the overall vision of the prototype design. The target wearers of the prototype are the family trio, namely the father, mother and child's three pieces of clothing for one parent-child clothing series, with the interactive effect that is transcending the traditional parent-child clothing.

The second one is the appearance design of the prototype. Interactive clothing is a future trend, so the style of the clothing and the selection of the basic fabrics to be more avant-garde.

Finally, it is the technical design of the interaction between physical signals and social symbols. From a psychological point of view, physical space distance can represent a social symbol of intimacy or alienation of interpersonal relationships (King O'Riain, 2016). Therefore, this study uses physical space distance as the signal trigger for interpersonal interaction. The spatial distance between these clothing was set as a physical signal of the interpersonal relationship, and the external appearance change of the clothing caused by the change of the spatial distance between these clothes is set as a social symbol which represents the change of the interpersonal relationship. Assuming that the three pieces of clothing will form a variety of interactions with one another, the specific design ideas, that is, the initial arrangement of time, task, and scheduling is as follows:

- When the distance between child and father two individuals wearing the prototypes was reduced, their clothing appearance produces a pattern of change at the same time.
- When the distance between the child and mother was reduced, their clothing produces a change simultaneously.
- When the child and parents were close to the set distance at the same time, three pieces of clothing simultaneously causing the change.
- When father and mother were closed, their clothing also causes a kind of change.
- Also, every piece of clothing has its own self-reaction.

4.3.3.3 Material selection
• Selection of new fabrics

Regarding clothing fabric, the inner fabric is mainly selected DuPont Tyvek paper fabric and 3M luminous reflective fabric splicing. The new type of DuPont paper fabric is 100% high-density polyethylene. Visually, it is presented in the form of paper, two-way ventilated, and corrosion resistance. After completing combustion, the fabric is left with only water vapour and carbon dioxide, which does not damage the environment at all. The luminous reflective fabric could show different colours in the state of light and no light, which has changed the traditional visual concept of clothing. The outer fabric of clothing is TPU Transparent film, also known as thermoplastic polyurethane, is a class of burning furnace air pollution-free fabrics. The smooth and transparent texture could show the fashionable atmosphere of high technology and futuristic.

- **Selection of network components and LED strip material**

Figure 4-10 is the pattern design of the three pieces of clothing, there are the father's clothing, mother's clothing and children's clothing from left to right. Then take the mother's clothing as an example to illustrate the hardware embedded content.

To create a network with clothing as an interactive medium, the basic network components embedded in the clothing include power supplies, sensors, chips, and microprocessors. In terms of electronic components, each piece of clothing is independently using a battery (positioned in ① in Figure 4-10) and a microcontroller STC51 (Jiaxing Xinwei Electronic Technology Co., Ltd.) (positioned in ② in Figure 4-10) as the main control chip to control the sensor response, LED and the communication of three pieces of clothing. In the communication node, the ARM Cortex-M4 microprocessor was adopted as the primary control chip. For the sensor selection, the horizontal KC_IRS (Xinyujia Electronic Technology Co., Ltd.) (Positioned in ③ in Figure 4-10) was adopted as an infrared sensor module.

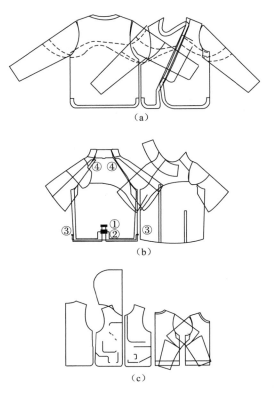

Figure 4-10　Pattern design and hardware embedding diagram

For the clothing-mediated manifestation of the social symbol output, two types of flexible light strips were selected to be embedded in the clothing. The LED using FPC as the base plate can be arbitrarily curved flexible flat strip with a 5050-60GRB model (Shanghai Jingzheng Lighting lamp Co., Ltd.) (positioned in ④ in Figure 4-10), can be fixed to a variety of letters or patterns of shapes. The EL cold light strip, with characteristics of repeatedly being folded and curved, low power consumption, and can achieve steady-slow, flash-fast and flash three kinds of effects, suitable for use in clothing design. The circuit design is shown in Figure 4-11.

4.3.3.4　Prototyping

The design and production details of this series of prototypes are shown in Figure 4-12. These prototypes' reflective fabric material and white, orange, black three colours combination embody the sense of science and technology due to their strong visual contrast and brightness contrast.

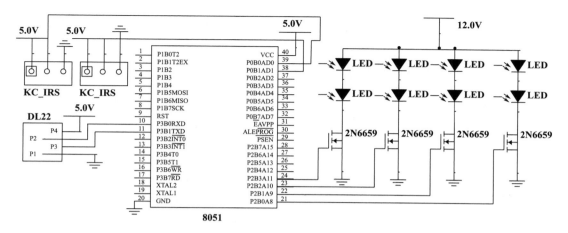

Figure 4-11   Circuit design

The colours and styles of the three-piece clothing are echoed, full of interest and relevance. Mother's clothing with different colour stitching with a large area of white, stable and no lack of lively, trumpet sleeves style enhanced stereoscopic and layered sense. Children's clothing is an orange hooded vest with a white sweater, it has a rhythm effect, and elbow's zipper segmentation enhances the profile of the fashion sense. The strong colour contrast between the left side and right side of the father's wear enhances the visual impact; the exaggerated curve of the overlap gives a fashionable atmosphere.

Figure 4-12   Photos of prototypes

- **Four types of interaction performances**

Note the special effects that are indicated by the white dotted circle in Figure 4-13.

(1) When the distance between child and father decreases within 2.5 meters, the two garments produce a reaction respectively. The performance form is the front of the children's wear LED thin-line strips and the father's clothing horizontal LEDs strips at the same time start shiny [Figure 4-13 (a)].

(2) When father and mother stand together within 2.5 meters, the heart-shaped LEDs pattern on the chest of father's clothing and the LEDs pattern of mother's shoulder shimmy simultaneously [Figure 4-13 (b)].

(3) When the distance between the child and the mother decreases within 2.5 meters, the two garments produce a reaction respectively. The manifestation is that the LED strips embedded in the hem of their garments start to shine at the same time [Figure 4-13 (c)].

(4) When the child and parents are close to the set distance within 2.5 meters at the same time, the sleeves of the three wears shine simultaneously produce a change [Figure 4-13 (d)].

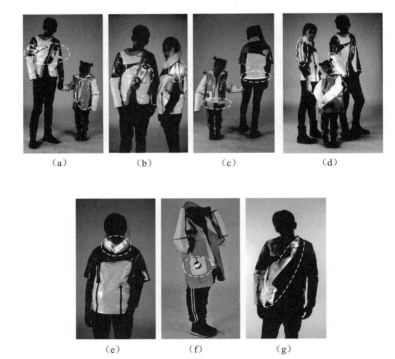

Figure 4-13 Interactive effects of prototypes

• **Three types of self-interaction performances**

(1) Mother's clothing, the embedded LEDs is activated when the collar buckled up, to achieve the expected interactive effect [Figure 4-13 (e)].

(2) Children's clothing, the reaction is through the Kcirs infrared sensor to the reaction of the hat, causing the front pocket decorative LED activated, to achieve the expected interactive effect [Figure 4-13 (f)].

(3) Father's clothing, its own reaction is triggered by the induction of the photosensitive

diode module, causing the front of the zipper part of the LED strips activated, to achieve the previous envisaged interactive effect [Figure 4-13 (g)].

According to the above, various demonstration effects of these prototypes, embedding CPS could greatly enrich the design concept and manifestation of the parent-child clothing. This series of clothing achieved by the interaction of individuals and groups as mentioned in the social level of the CPCS model, realizes the representation of the social symbol from the physical distance signal recognition to the parent-child relationship. This simple transformation process from data into information and knowledge illustrates the application forms of CPS in the process of interactive clothing design and development which response to the aim of this case study, MRQ and SRQ 1 mentioned above.

4.4 Summary

In this chapter, through the development of the infant's smart clothing prototypes in case study 1, the idea of the feasibility of IoT sensing applied to clothe is verified. The application effect of prototypes and the acquisition of national patents show that IoT technology can be applied to our daily clothing life, which lays a practical foundation of interactive clothing development for the follow-up study of the project.

Through the prototyping of interactive couple clothing in case study 2, the interactive reactions between two pieces of clothing were realized. Satisfactory prototyping results, validating the hypothesis that the development of multiple items of clothing worn by different individuals and having a mutual reaction relationship. By triggering design, interactive clothing can reflect some interpersonal and emotional relationship, such as happiness, anger, sadness, and joy between wearers. The satisfactory effect of the prototype also illustrates the reasons why smart clothing should satisfy the IoT technical functions and human emotional expression simultaneously. It also prompts us to investigate the manner in which artistic design perspectives and engineering methods combined effectively, to explore the R&D elements of future smart clothing based on IoT technology.

In case study 3, the diversified interaction results achieved through a three-piece set of parent-child clothing. The prototype development process successfully grafted the CPS architecture in the field of information science and art design in the field of humanistic science from an interdisciplinary perspective for interactive fashion innovation, enriching the expression form of social symbol semantic of clothing. By using the CPCS architecture model in the hyper world, and takes the development process of interactive parent-child clothing as a case, to analyze the transformation from the physical signal input to the social symbol recognition output. Four types of interactive performances and three types of self-interaction performances have verified the feasibility of the CPCS model adopted in interactive clothing design.

Chapter 5
Emotional Evaluation, Data Analysis, and Discussion

5.1 Data analysis and evaluation on interactive couple clothing

5.1.1 Kansei evaluation I

We used the semantic differential (SD) method to analyze the effects of the interactive clothing on the psychological responses of wearers. That is, the participants are required to analyze the smart clothing as wearers. Several emotion models exist. For example, Larsen and Diener (1992) have developed a model divided into eight categories (Figure 5-1a-h), which can be applied to both active and inactive smart clothing. We employed this eight-way split model as an evaluation reference to consider the relationship between the emotions of the wearers and the clothing worn by various couples. Each emotion was quantified as a value within the range of 1-10 by using fuzzy inference.

We conducted an emotional survey evaluation experiment under activating and inactivate range sensors conditions for the LED sensors embedded in each couple's clothes. Considering the future main target consumers of smart clothing are expected to be college students and other young people, 34 boys and girls college couples participated in this wearing experiments. 16 people of them in the laboratory light in a darker environment wearing experience, with a time of about 30 minutes; 18 people in the outdoor night environment, and the experience of wearing for about 60 minutes.

Each group of participants was a boy and a girl with a couple of relationships. We require each pair of wearers in the course of the experiment as much as possible in the daily life of the behavior to act, and to complete the basic process of two people approaching from more than 5 meters and then gradually pulling away.

The form of filling out questionnaires after wearing experience was adopted. The experiment was divided into two steps. In the first group of tests, each pair of couples in the clothing without activating the sensor case of wearing experience and fill out the emotional value test questionnaire. In the second group of tests, each couple's wearing experience is carried out in the case of sensors activating and then rate the emotion scale.

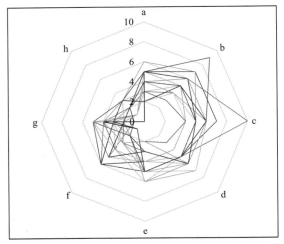 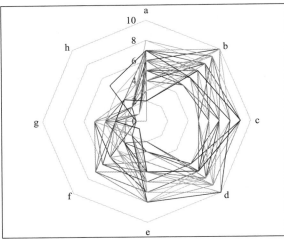

Figure 5-1　Results of Kansei evaluation on emotional expression for non-interactive and interactive, for eight-way split model

Notes: (a) arousal; (b) excitement; (c) joy; (d) fun; (e) composure; (f) laziness; (g) sadness; (h) anger.

The survey data for the active and non-interactive clothing differed significantly. In the case of the inactivate scale for the LED sensor, eight-way split model in Figure 5-1 shows that the main range of the evaluation index fluctuation is within 6. However, for the activate range of the LED sensor, Figure 5-1 shows that the range of the evaluation index fluctuates obviously, with many of the evaluation values reaching the maximum of 10. The average emotion values for the interactive and non-interactive are shown in Table 5-1. The paired samples t-test revealed that $P=0.008 \leqslant 0.05$. From the results of this Kansei evaluation, it can stimulate the emotional response of the participants to achieve a higher score in the activate sensor state; therefore, H1 and H2 mentioned in Chapter 4 are validated.

Table 5-1　Differences in the evaluation of average values

Status	a	b	c	d	e	f	g	h
Non-interactive	4.00	5.26	5.59	5.47	4.47	3.50	2.91	1.29
Interactive	5.09	7.82	6.85	8.29	5.62	3.94	3.35	1.50

Notes: Average emotion values for interactive and non-interactive, for eight-way split model.

5.1.2　Kansei evaluation Ⅱ

After completing the emotional value evaluation, follow-up comparative evaluation experiment by observation method is continued. That is, the participants are required to analyze and judge the differences between various types of smart clothing as bystanders.

Chapter 5
Emotional Evaluation, Data Analysis, and Discussion

• **Experimental form**

Multi-category smart clothing comparative evaluation by observation method. For the corresponding clothing pictures and video using the observation evaluation method. The evaluation score is a scale comprised of seven ranks on a questionnaire sheet. Participants responded to each adjective scale by marking the rating after watching the specified smart clothing videos and pictures.

• **Evaluation subject**

The college students who have participated in previous experiments. Including design disciplines (as the designer cluster), information engineering disciplines (as engineer cluster) and other disciplines (as consumer cluster) of the three clusters, each of 6 people. Because smart clothing design involves the integration of art design and information sensing technology, as well as the need for non-professionals to the comprehensive evaluation of the consumer's identity, the three clusters of participants, are selected for comparative evaluation. These participants had participated in the first experiment, and a return visit was inviting them to participants in the second experiment was conducive to the continuity of the experiment.

• **Evaluation object**

Prototype [Figure 5-2 (a)], ANREALAGE [Figure 5-2 (b)], smart sportswear [Figure 5-2 (c)] (De Acutis and De Rossi, 2017), healthcare clothing [Figure 5-2 (d)] (Chen et al., 2017).

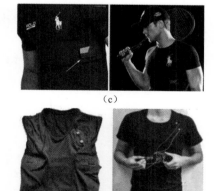

(a)　　　　　　　　(b)　　　　　　　　(c)
　　　　　　　　　　　　　　　　　　　(d)

Figure 5-2　Evaluation object

Annotation of compare objects:
- The ANREALAGE [Figure 5-2 (b)] is a Japanese fashion brand. Its designer, Kunihiko Morinaga, is known as the "Revolutionary of the Digital Age", specializing in the

combination of clothing and technology, he has endless innovation and fantasies about fabrics.

- Ralph Lauren OMsignal Polo Tech shirt is a typical smart sportswear [Figure 5-2 (c)]. It is the first luxury sports lifestyle brand offers smart apparel collection and tested in the 2014 US Open. This shirt uses biometric technology to collect athlete's physiological signals including heartbeat and respiration as well as some psychometrics.
- Wearable 2.0 healthcare system [Figure 5-2 (d)], which consists of sensors, electrodes, and wires, is the critical component to collect users' physiological data and receive the analysis results of users' health status provided by cloud-based intelligence.

• **Evaluation words**

By referring to the "Human aspects in smart clothing were dealt with usability, functionality, durability, safety, comfort, and fashion" proposed by Cho, et al., (2009), and the consumer needs of functionality, expressiveness, and aesthetics (FEA) raised by Hwang et al., (2016), the evaluation vocabulary was identified as functional, practical, futuristic, emotional transmission, appearance style and interactive technology.

• **Evaluation scales**

Take item "functional" as example, we divided "functional" into semantic differential scales with seven points from the lowest level 1 of "functional" to the highest level 7.

Overall, Figure 5-3 shows that the average score of the prototype is significantly higher than other three types of smart clothing in the three clusters' evaluation of these four types of smart clothing. Since the evaluation of the prototype is the highest, then it is necessary for us to extract three clusters to analyze the details of the prototype evaluation separately.

Figure 5-4 uncovered designer cluster thinks the best point of the prototype is future-style, emotional and interactive; the most vulnerable are functional. Figure 5-5 reveals the engineers' view that the prototype's best point is future-style and emotional, the weakest point being functional. Figure 5-6 shows that consumer cluster thinks the best point of the prototype is future-style, the most vulnerable are functional and practical.

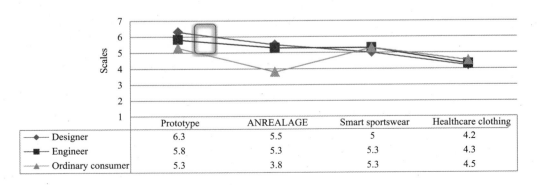

Figure 5-3 Comparison of average values by designer, information engineer and consumer clusters

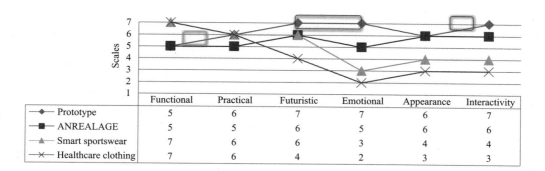

Figure 5-4 The evaluation value of designer cluster to four kinds of smart clothing

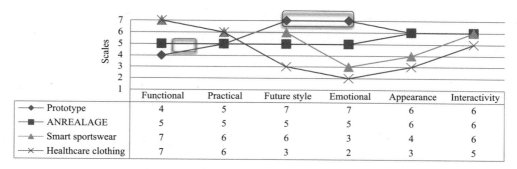

Figure 5-5 The evaluation value of information engineer cluster to four kinds of smart clothing

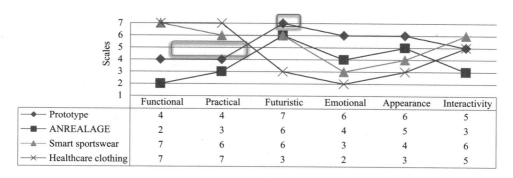

Figure 5-6 The evaluation value of ordinary consumer cluster to four kinds of smart clothing

Comprehensive evaluation of different groups, people's expectations of the future of smart clothing are not just focused on the use of high-tech. The three clusters have the highest degree of satisfaction with the futuristic and emotional elements of the prototype. Therefore, the idea proposed by H3 that the emotional interaction is the main direction of future smart clothing development is established.

However, the functional satisfaction of the three clusters to the prototype is not higher, which indicates that the balance between the application of high-tech and emotional interaction in the product design process need to improve. We can see that the design concept of implanting wearable technology in clothing to improve the function of garments is not yet a widespread resonance among consumers. Only by implantable smart technology to effectively improve the products more in line with the needs of emotional communication, more in line with the expected lifestyle, these are the most important indicators for people to decide whether to wear smart clothing. This is the key content of future smart clothing design, but also the "WHAT" element mentioned above.

5.1.3 Case discussion

Human-centered and emotion-based design for "WHAT" element and H3 mentioned above. It is necessary to provide clothing with vitality rather than, simply, embedded IoT technology, if the clothing is to express the wearer's emotions vividly. Although this is a difficult task, deep-thinking designers to identify future subjects should use the socially innovative design perspective (Nagai, 2015). Thus, it is crucial for forward-thinking designers to embrace this challenge.

The design process for interactive clothing should be human-centered, rather than focusing on technology. Interpersonal communication is the focus of social existence, and for useful technology must support socialization (Tao, 2005). Nelson and Stolterman (2012) offered a formulation of design culture's fundamental ideas, in the design process applied to an infinite variety of design domains. Further, Giacomin (2014) has stated that the model of human-centered design is based on a pyramid hierarchy, in which interactive, sociological considerations and the metaphysical meaning contact with the design. Because clothing has a metaphysical attribute, Baurley (2004) has proposed that people can interact with the clothing of others nearby by changing the visual appearance. For the design process, Schütte (2005) and Lee et al. (2002) contributed to the balance between functional technology and emotional expressiveness.

The human-centered approach cannot neglect consideration of human emotions. For product design, traditional cognitive approaches have tended to underestimate the user emotions which act as a critical component of artifact sense-making (Rafaeli, and Vilnai-Yavetz, 2003; Spillers, 2004). Wensveen (2002) has designed a mood alarm clock that illustrates the importance of a tight coupling between the emotional level of interaction, and the actual use of such interactive technology.

As shown in Figure 5-7, we should further investigate the structure of the emotions underlying human clothing-related behavior, and obtain a more inclusive vision of the social psychology of smart clothing. Interactive clothing must satisfy both technical functionality and human emotional expression requirements (Bakker and Niemantsverdriet, 2016). In other words, both basic and advanced social properties of clothing must be incorporated. These also correspond to the third design factor "WHAT" and H3 introduced above.

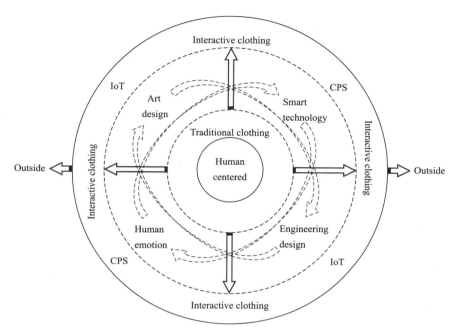

Figure 5-7　Human-centered R&D process for interactive clothing

In this case study, interactive clothing for couples was examined using experimental prototypes. The efficacy of this clothing was then assessed using a Kansei evaluation. In hence, it was found that the prototypes satisfied the criteria of stage 2.0 of the smart clothing (Figure 2-6) and the interactive clothing evolutionary roadmap (Figure 2-13).

For interactive fashion designers, design for the future IoC, rather than focusing on purely functional behavioral or aesthetic-appeal criteria, emphasis should be placed on blending technological developments in engineering with emotional responses in product design, on identifying artifacts that trigger and mediate emotional responses, and on seeking the emotional structure underlying human clothing-related behavior.

Artistic design perspectives should be effectively combined with engineering methods during the R&D of interactive fashion products. The primary task of design thinking is to achieve the external aesthetic form without excessive consideration of the technical characteristics of the product. In contrast, the engineering approach aims to realize technological applications without emphasizing the added value or social attributes of the product. An effective designer should embrace the opportunities and challenges of achieving smart clothing evolution from stage 1.0 to 3.0. Such a designer should exhibit both artistic creativity and engineering rigor, integrating design and engineering perspectives, and blending emotional responses with smart technology to achieve interactive fashion innovations.

5.1.4 Limitations and further research

The sensors combined with LED to display technologies is a relatively mature technology, the use of this technology in prototype design simply for the convenience and easiness to achieve an interactive effect, so the project team did not develop special sensing technology. The interactive effects and comfort (wearability) of the prototype should be further improved if specialized, advanced, washable, durable, and flexible sensing elements or smart fabrics are developed in the future. During the experiment, the participants' wearing experience time is limited, and the evaluation is mainly subjective. The evaluators participates are all college students, so the evaluation results may not be universally representative.

In the next research, we will focus on the hierarchy and typology of smart clothing R&D based on humanities and technology perspective simultaneously, especially focus on the new concept of IoC which we have pioneered proposed.

5.2 Data analysis and evaluation on interactive parent–child clothing

According to the CPCS model, in the process of interactive clothing design, the formation of wisdom needs to reach the cognition level through knowledge mining, i. e., humane evaluation. Because this case study uses parent-child clothing that emphasizes emotion as research carrier, so the evaluation approach adopts the Kansei Engineering method rather than ergonomic or functional evaluation methods. By comparing the prototype with the ordinary parent-child clothing, the advantages and disadvantages of the prototype design are analyzed, and the groundwork for the subsequent configuration technology flow is laid.

5.2.1 Selection of comparative evaluation objects

Evaluation elements include casual style, formal style, graffiti style, deconstruction style, national style and our prototype, i. e., optical sensing interactive style (Figure 5-8).

These six categories could cover and represent the existing types of parent-child clothing on the market as follows.

(a) **Casual style**

This style of clothing refers to people wearing in an unrestrained and free leisure life, emphasizing a relaxed and straightforward style.

According to Wikipedia, casual style wear *is a western dress code category that*

Chapter 5
Emotional Evaluation, Data Analysis, and Discussion

Figure 5-8　Photos of comparative categories

comprises anything not traditionally appropriate with more formal dress codes: formal wear, semi-formal wear, or informal wear. In general, casual wear is associated with emphasizing personal comfort and individuality over formality or conformity. As such, it may referred to as leisurewear. In a broader sense, the word "casual" may be defined as anything relaxed, occasional, spontaneous, "suited for everyday use", or "informal" in the sense of "not formal" (although notably informal attire actually traditionally refers to a Western dress code more formal than casual attire, a step below semi-formal attire).

Source: Official website of Wikipedia

(b) **Formal style**

The design of this style clothing is concise, elegant and dignified. It highlights the texture of the fabric and the fine cutting and shows a classical beauty.

According to Wikipedia, formal wear *is the traditional Western dress code category applicable for the most formal occasions, such as weddings, christenings, confirmations, funerals, Easter and Christmas traditions, in addition to certain audiences, balls, and horse racing events. Generally permitted other alternatives are the most formal versions of ceremonial dresses (including court dresses, diplomatic uniforms and academic dresses), full dress uniforms, religious clothing, and most rarely frock coats. In addition, formal attire may be instructed to be worn with official orders and medals.*

Source: Official website of Wikipedia

(c) Graffiti style

This style refers to the various patterns of clothing contrast strong, color blending. An extraordinary, exaggerated or casual style depicts the graphic, text or color in an abstract sense.

(d) Deconstructive style

This style emphasizes the concept of unstructured design. It uses the concept of decomposition, emphasizing breaking, superimposing, reorganizing, attaching importance to the individual or the component itself, and opposing the overall unification to create fragmentation and uncertainty feeling.

(e) Interactive style

This style refers to smart clothing that developed using Cyber-Physical Systems /Internet of Things technology. The distance, sound, movement between the pluralities of clothes (dressers) can automatically adjust the appearance, colour, model of the clothing, and the smart interactive reaction is formed between many pieces of clothing.

(f) National style

This style refers to the clothing that combines the elements of traditional national costumes, combined with the needs of modern life, social and other occasions, with both national elements and modern fashion design elements. This costume shows the ultimate symbol of national culture, and its design elements include, for example, Japanese kimonos, Mexican jeans, Middle Eastern keffiyeh, Scottish kilts, and so on.

5.2.2　Selection of semantic opposite adjective

Three categories of words, total 24 pairs of antonym phrases were constructed, respectively choice from the three-level concept of parent-child clothing, which was summarized by the prior interview: the clothing style design, engineering technology, parent-child clothing extension meaning (Table 5-2).

Table 5-2　Semantic opposite adjective pairs

	Category Ⅰ: Style design level			Category Ⅱ: Technology content level			Category Ⅲ: Extended meaning level	
No.	Positive	Negative	No.	Positive	Negative	No.	Positive	Negative
1	Avant-garde	Conservative	10	Smart	Rigid	19	Affinity	Alienated
2	Staid	Lively	11	Environmental	Destructive	20	Warm	Cold
3	Exquisite	Rough	12	Flexible	Bound	21	Interesting	Stodgy
4	Childlike	Mature	13	Technology	Traditional	22	Surprise	Disappointed

Continued

Category I: Style design level			Category II: Technology content level			Category III: Extended meaning level		
No.	Positive	Negative	No.	Positive	Negative	No.	Positive	Negative
5	Creative	Monotonous	14	Interactive	Isolated	23	Relaxed	Gloomy
6	Elegant	Vulgar	15	Open	Restricted	24	Enthusiastic	Indifferent
7	Fashionable	Outdated	16	Systematic	Messy			
8	Cross-border	Closed	17	Multifunctional	Single			
9	Concise	Tedious	18	Emotional	Rational			

5.2.3 Participants' selection and evaluation methods

Clothing industry personnel, information science or optoelectronics engineering personnel, and children's parents. The reasons for choosing these three groups as participants are as follows: Clothing industry personnel, specially engaged in clothing design professionals, could compare clothing to make a professional evaluation. Because the prototype of this experiment used information sensing technology and optoelectronic technology, therefore, it is hoped that the researchers of information engineering and optoelectronic engineering could give the corresponding evaluation from the perspective of the application of intelligent technology. The target purchase group of parent-child clothing is the parents' cluster so that the result will be more convincing with the help of the target users' evaluation.

The participants, including 14 people engaged in the clothing industry (6 male, 8 female; 7 participants is 18-29 years old, 1 participant is 30-39 years old, 6 participants is 40-49 years old), 23 photoelectric information technicians (11 male, 12 female; 16 participants is 18-29 years old, 5 participants is 30-39 years old, 1 participant is more than 40-49 years old, 1 participant is 50 years old) and 36 children's parents (7 male, 29 female; 1 participant is 18-29 years old, 30 participants is 30-39 years old, 5 participant is more than 40-49 years old).

In particular, it should be emphasized that the questionnaire in the clothing evaluation link style 1 to style 6 of the question is the same 24 questions, in order to ensure the credibility of the results of the questionnaire feedback, we randomly disrupted the sequencing of these questions.

After providing brief introductions to participants about the different contexts of the six categories evaluate elements, we ask participants according to their knowledge background and living habits to rate each of perceptions regarding these six categories parent-child clothes used semantic differential scales by observation method. For the evaluation scales, take pair item "Smart and Rigid" as an example (Table 5-3), we divided into 7 points from the "very smart" level to the "very rigid" level. Finally, 73 valid questionnaires were collected.

Table 5-3 Example of an SD scale with seven points for adjective pair

Very smart	Quite smart	Slightly smart	Neutral	Slightly rigid	Quite rigid	Very rigid
3	2	1	0	−1	−2	−3

Before statistical data, we first tested the rationality of the selection of participants in this differentiated group with the stacked column chart. According to the questionnaire scoring statistics of designers, engineers and parents, we can find that the peak values of the designer cluster are between 60 and 80 (Figure 5-9), the peak of the engineering cluster is between 100 and 120 (Figure 5-10), and the peak of the parent cluster is between 160 and 180 (Figure 5-11). Due to the significant differences in cluster evaluation scores, the selection of participants using this differentiated cluster could provide a guarantee of the credibility of the evaluation.

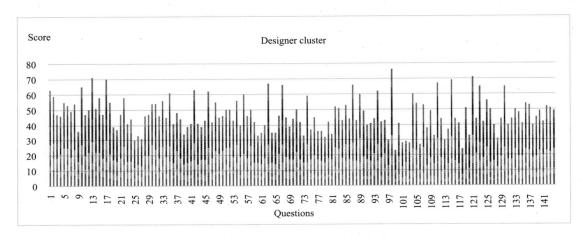

Figure 5-9 Evaluation value of 14 designers

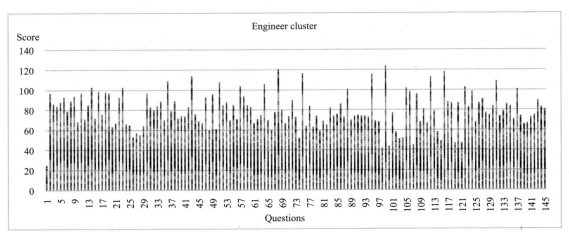

Figure 5-10 Evaluation value of 23 engineers

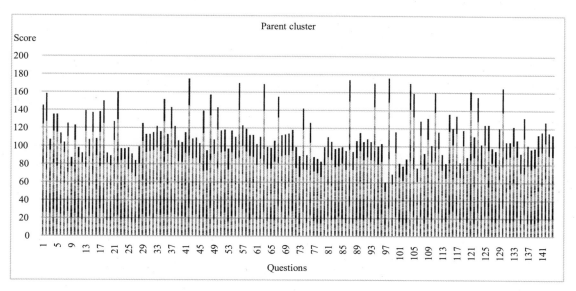

Figure 5-11 Evaluation value of 36 parents

5.2.4 Differences between the two categories of parent-child clothing

According to the scale statistics on the average value of the evaluation, the comparison between the prototype and other five categories of traditional parent-child clothing is made to excavate the difference between the interactive and the traditional parent-child clothing.

According to the 73 participants' choice of "desire to buy" in all six categories of parent-child clothing, the highest ranking is the prototype, i. e. Optical sensing sci-fi style (accounting for 32.9%), followed by the Dress style (26%) which is also the highest ranking among the five categories traditional parent-child clothing. Therefore, we chose the Dress style as an object compared to the prototype.

According to the results of the paired-samples t-test between prototype and traditional parent-child clothing (Table 5-4), 16 of the 24 constructs have significant fixed effects with a p-value lower than 0.05. The creativity and other indicators of the prototype are significantly higher than that of the traditional parent-child clothing. However, the disadvantages of the prototype are just the emotional extension elements. There is the most significant difference between the bold and underlined parts of the t-value in Table 5-4. The constructs corresponding to these values, especially the Concise, Elegant, Systematic, Exquisite and Warm options, are the weakest point of the prototype, but also the key area for future development that needs to be improved. The Smart, Technology, Cross-border, Multifunctional, and Interactive can represent the advantages and characteristics of the prototype.

Table 5-4 Results of the paired-samples *t*-test

Style design level			Technology content level			Extended meaning level		
Constructs	Prototype	Traditional	Constructs	Prototype	Traditional	Constructs	Prototype	Traditional
Avant-garde	39.7%	19.2%	Smart	45.2%	1.4%	Affinity	21.9%	16.4%
Staid	5.5%	13.7%	Environmental	24.7%	12.3%	Warm	21.9%	20.5%
Exquisite	17.8%	23.3%	Flexible	34.2%	13.7%	Interesting	23.3%	16.4%
Childlike	13.7%	9.6%	Technology	45.2%	1.4%	Surprise	17.8%	12.3%
Creative	43.8%	12.3%	Interactive	34.2%	5.5%	Relaxed	15.1%	15.1%
Elegant	11%	24.7%	Open	17.8%	5.5%	Enthusiastic	24.7%	19.2%
Fashionable	23.3%	30.1%	Systematic	4.1%	12.3%			
Cross-border	47.9%	8.2%	Multifunctional	39.7%	8.2%			
Concise	5.5%	11%	Emotional	21.9%	15.1%			

To ensure the rationality of the analysis, we continued the frequency statistics method. We separately extracted the five semantic adjective items with the highest frequency value of the above-mentioned two categories of clothing evaluation as the index of differential analysis (Table 5-5). These five indicators are also the characteristics and advantages of the two categories of parent-child clothing respectively. According to the value of the frequency statistic scale, the highest score of the prototype is Cross-border (47.9%), Smart (45.2%), Technology (45.2%), Creative (43.8%) and Multifunctional (39.7%). While the advantage of the Dress style clothing is Fashionable (30.1%), Elegant (24.7%), Exquisite (23.3%), Warm (20.5%) and Enthusiastic (19.2%), they are also the key improvement factors of future interactive parent-child clothing development.

Table 5-5 Top five value in prototype and traditional parent-child clothing

Contrastive semantic		Mean	SD	SE	*t*	*P*
Rigid	Smart	1.781	1.557	0.182	**9.774**	0.000
Monotonous	Creative	0.890	1.663	0.195	4.575	0.000
Outdated	Fashionable	0.027	1.624	0.190	0.144	0.886
Closed	Cross-border	1.411	1.373	0.161	**8.782**	0.000
Vulgar	Elegant	−1.027	1.972	0.231	**−4.452**	0.000
Tedious	Concise	−1.466	2.109	0.247	**−5.939**	0.000
Destructive	Environmental	0.178	1.743	0.204	0.873	0.386
Bound	Flexible	1.068	1.719	0.201	5.312	0.000

Continued

Contrastive semantic		Mean	SD	SE	t	P
Traditional	Technology	1.781	1.669	0.195	**9.118**	0.000
Isolated	Interactive	1.192	1.560	0.183	**6.526**	0.000
Open	Restricted	−0.137	2.057	0.241	−0.569	0.571
Messy	Systematic	−1.082	2.228	0.261	**−4.149**	0.000
Single	Multifunctional	1.808	1.912	0.224	**8.079**	0.000
Rational	Emotional	−0.616	2.271	0.266	−2.319	0.023
Alienated	Affinity	0.219	1.652	0.193	1.134	0.261
Cold	Warm	−0.603	1.614	0.189	**−3.191**	0.002
Stodgy	Interesting	−0.096	1.108	0.130	−0.740	0.462
Disappointed	Surprise	0.164	1.871	0.219	0.751	0.455
Gloomy	Relaxed	−0.192	1.883	0.220	−0.870	0.387
Indifferent	Enthusiastic	0.521	1.519	0.178	2.927	0.005
Conservative	Avant-garde	0.630	1.568	0.184	3.434	0.001
Lively	Staid	−0.397	2.559	0.300	−1.326	0.189
Rough	Exquisite	−0.753	1.949	0.228	**−3.302**	0.001
Mature	Childlike	0.630	2.010	0.235	2.678	0.009

Just as Apple CEO Tim Cook delivered MIT's 2017 commencement speech: technology should marry with the liberal arts and humanities that makes our hearts sing. Researchers should take the human body as a critical frontier to create powering and deeply engaging ways of interacting with computers (Broadhurst, & Price, 2017). For clothing that embodies people's feelings, embedding CPS can easily achieve interpersonal interaction effect. Nevertheless, the results from our evaluation experiment show that technology-oriented clothing is relatively ruthless and lacks humanistic emotion. Therefore, how to enhance the emotional value of smart parent-child clothing should be one of our next research goals.

5.2.5 Limitations and follow-up studies

In the application of information technology, the prototype developed in this case study is limited to the category of interactive manifestation and does not involve the scope of the big data measure and algorithm analysis of CPS (Kindness et al., 2017). For the challenges mentioned in the introduction of this case study, that is, the differential needs of different objects, we need to deepen the research of machine learning and optimize the algorithm of merging time, task, and scheduling. The mining and categorization of the basic physical signals and physiological signals of the wearer, as well as the recognition and classification of the social symbols that can be displayed by the clothing, will help to improve the functional development and expand the application of the interactive clothing.

In the context of product design and evaluation, the limitations of the comparative method of image observation and evaluation may lead to the deviation of the experimental result from the real situation. Clothing is a kind of product that not only pays attention to the visual feeling but also needs to wear experience. The interactive parent-child clothing emphasizes the interactive experience. Of all those who took part in the survey, only 2 were from Japan and the remaining 71 were from China. Since participants were too concentrated in a single country, the data in this survey had some limitations. Therefore, the follow-up study will adopt the evaluation method of participants' wearing experience test, and participants from different countries should be invited to participate in the experiment.

5.2.6 Case conclusion

This case study summarizes the relationship between CPS architecture and interactive clothing design from a transdisciplinary perspective and takes the development process of interactive parent-child clothing as a case to verify the feasibility of the CPCS model for interactive clothing design.

This case study provides a fruitful practical application reference for non-information engineers, especially for clothing designers who are engaged in the art design field but not familiar with the relevant information science and technology, and explains the application principle and technical process of CPCS in the process of interactive clothing development. The satisfactory results of the design and evaluation of prototypes demonstrate that the potential of HCI combined with CPS for developing interactive clothing has just begun to be tapped, and there are more technical methods and opportunities than ever before to create compelling everyday dress experiences and deep personalization for fashion pioneers.

5.3 Summary

Through three progressive prototyping, evaluation, analysis and discussion, the feasibility of the CPCS model has been verified.

In the case study 2 for couple clothing, the evaluation Ⅰ indicated the prototype can stimulate the emotional response of the participants to achieve a higher score in the activate sensor state, therefore, the emotional interaction and transmission between people or between people and artifact may be more critical to smart clothing's success than its practical function elements. The evaluation Ⅱ revealed that in the process of interactive clothing design, the technical functionality should synchronize with the requirements of human emotional expression. Therefore, we proposed that the interactive fashion designers should design for the future IoC, rather than focusing on purely functional behavioral or aesthetic-appeal criteria, emphasis should be placed on blending technological developments in engineering with emotional responses in clothing design, on identifying artifacts that trigger and mediate emotional responses, and on seeking the emotional structure underlying human clothing-related behavior. Artistic design perspectives should be effectively combined with engineering methods during the R&D of interactive fashion products.

In the case study 3 for parent-child clothing, the evaluation results clarify the advantages and disadvantages of interactive clothing, and also bring enlightenment to the social semantic interaction symbol and form excavation of interactive clothing.

Based on the scale statistics on the average value of the evaluation from clothing industry personnel, information science or optoelectronics engineering personnel, and children's parents, the comparison between the prototype and the other five categories of traditional parent-child clothing is made to excavate the difference between the interactive and the traditional parent-child clothing. The score with significant differences illustrates that the emergence of successive waves of CPS can enable new symbolic interactive ways of coupling the clothes to different wearers. Besides, according to the value of statistic scale, the advantage of the prototype is Cross-border, Smart, Technology, Creative and Multifunctional, while the disadvantage is Concise, Elegant, Exquisite, Warm and Enthusiastic which are the critical improvement factors of interactive parent-child clothing development. From interactive representation, it is necessary to meet the differentiated needs of different types of wearer individuals or groups because of the interaction among different wearer individuals or groups, as well as the differences in various issues and attributes among each wearer or group. Different types of feedback forms are represented by different clothing external symbols. Various communication symbols that can express social message to be output during the interaction of different individuals or groups.

Chapter 6
Case Study: Purchase Intention and Design Elements of Parent-child Clothing

6.1 Case introduction

Through the analysis of consumers' attitude (ATT) and purchase intention (PI) towards parent-child smart clothing, this case study reversely derives and extracts the design elements of this type of clothing. This case study expands the category of Technology Acceptance Model (TAM) with clothing design attribute.

Based on perceived usefulness (PU), perceived ease of use (PE) and perceived performance risk (PR), functionality (FUN), aesthetic (AES) and compatibility (COM) of clothing are added to analyze the factors affecting consumers' ATT and PI towards parent-child smart clothing. A total of 372 volunteers participated in the test, and the results show that COM has a significant positive influence on PU, PE and PR. PU and FUN have a positive influence on purchase ATT and PI. PE positively affects PU and positively affects purchase ATT. PE positively affects PU and positively affects purchase ATT. AES positively influences purchase ATT but has little impact on PI. PR negatively influences both purchase ATT and PI but have little impact on PI.

This case study confirms the significance of multi-dimensional features of smart parent-child clothing, extracts and preliminarily establishes the framework model of design evaluation elements. And the results are helpful to the product design and development of parent-child smart clothing in the future.

6.2 Research model and hypotheses

TAM is considered as the most effective model to explain the acceptance level and use intention of information technology (Davis, 1989; Venkatesh, 2000). This model explains consumers' acceptance ATT and behavior towards innovative technologies such as information technology.

Among which, perceived usefulness (PU), perceived ease of use (PE), perceived performance risk (PR) and other factors have been considered to be the most recognized factor influencing consumers' ATT towards technology and innovative products and PI. At present, TAM has been widely used to evaluate consumers' acceptance of technology-related elements in the fashion industry. Such as the researches on the application of smart storage technology in the retail industry and consumers' acceptance of solar smart clothing (Hwang, 2016; Kim, 2017). Therefore, this study extends the theoretical analysis scope of TAM, adds functionality (FUN), aesthetic (AES), and compatibility (COM) of the clothing based on PU, PE, and PR, and analyzes factors affecting consumers' ATT and PI towards parent-child smart clothing.

A total of 14 hypotheses were proposed in this study. FUN, COM and AES are used as independent variables. PI is dependent variable, PU, PE, PR and ATT are mutually independent variables and dependent variables. Figure 6-1 is the research model presented in the form of structural equation model (SEM).

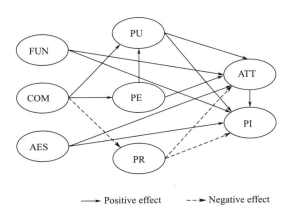

Figure 6-1　Research model

6.2.1　PU

PU is defined as "a person believes that a particular system can be used to improve the level of his or her job performance" (Davis, 1989). Venkatesh believes that the usefulness of a product directly affects their ATT towards technology adoption, and PU has been regarded as the most powerful predictor of willingness to use and adopt technology, and a key variable that affects consumer ATT (Venkatesh, 2003). Chanmi and Chae et al. consider that PU gas a positive impact on purchase ATT and PI of solar clothing. 3 Based on the above research results, it shows that PU positively influences ATT and PI of parent-child smart clothing:

- *Hypothesis 1a：PU positively influences ATT of parent-child smart clothing.*
- *Hypothesis 1b：PU positively influences PI of parent-child smart clothing.*

6.2.2 PE

PE refers to "the degree to which a person thinks it is effortless to use a particular system", in a number of relevant researches, for example, Kim et al. propose that PE is a key factor in determining the adoption of smart retail technology and positively influences consumers' ATT towards wearable fit index technology (Lunney, 2016), PE will positively influence PU. However, Ko et al. consider that the relationship between the complexity and PI of smart clothing is not significant. Therefore, it is expected that, PE positively influences PU and ATT of solar clothing:

- *Hypothesis 2a: PE positively influences PU.*
- *Hypothesis 2b: PE positively influences purchase ATT of parent-child smart clothing.*

6.2.3 PR

PR is defined as "the uncertainty that consumers face when they cannot foresee the consequences of their purchase decisions". Dowling and Staelin (1994) also conceptualized as consumers' perception of the uncertainty and adverse consequences of purchasing a product or service. Chen and Grewal et al. believed that performance risk and financial risk are the most commonly used assessment risks among perceived risks (Chen, 2003; Grewal, 1994). Financial risk refers to the customer's possible loss in the currency, including product repair or replacement and refund. Performance risk is defined as the potential loss caused when a product fails to meet the expectations of consumers, including the risk of innovation. Simply put, the more innovative the product, the higher the uncertainty associated with this new feature, and the more likely consumers are to hesitate to buy such a product. In short, PR means that the more innovative the product is in terms of the clothing, the higher the uncertainty associated with this new feature will be, and the more likely consumers are to hesitate to purchase such products. Therefore, innovation-related uncertainty can be conceptualized as PR, which plays an important role in the formation of new product ATT and PI for these products. Generally, PR negatively influences ATT towards the adoption of innovative technologies, and relevant researches have also proposed that PR negatively influences PI (Park, 2005; Kaisa, 2018). Hence, we assume that the PR of parent-child smart clothing negatively influences ATT and PI:

- *Hypothesis 3a: PR negatively influences purchase ATT of smart parent-child clothing.*
- *Hypothesis 3b: PR negatively influences PI of smart parent-child clothing.*

6.2.4 FUN

In addition to the analysis of parent-child smart clothing as an innovative and technologically integrated product, the complex external attributes of clothing should also be discussed. Three dimensions play a crucial role in clothing design to meet consumers' demand for innovative design, namely, FUN, COM, and AES (Lamb, 1992).

FUN dimensions of clothing include FUN, protection and comfort, which are related to the practicability of clothing and affect the acceptance of technology by users. Because of the different needs of the audience, the multi-functional development of smart clothing is a necessary trend. At present, the relevant direction of parent-child smart clothing is mainly from the perspective of interaction design, health monitoring, and other multi-functions. Therefore, we believe that FUN will positively influence purchase ATT and PI of parent-child smart clothing.

- *Hypothesis 4a*: *FUN positively influences purchase ATT of parent-child smart clothing.*
- *Hypothesis 4b*: *FUN positively influences PI of parent-child smart clothing.*

6.2.5 COM

COM dimension refers to the symbolic communication characteristics of identity, such as values, roles, and self-esteem. Based on the socio-cultural and psychological aspects of clothing, the products should be compatible with the status and self-image of the wearer, which leads to the importance of COM (Stokes, 2012). The sense of innovation is also a kind of COM, which affects users' acceptance to smart clothing. 4 An innovative product can reduce the risk of technical failure if it is simple and uncomplicated to use. In short, if the product meets consumers' current needs and lifestyle, it will have a positive impact on users' technology acceptance. Therefore, we suggest that COM positively influences technology acceptance variant of PU and PE, and negatively influences PR:

- *Hypothesis 5a*: *COM positively influences PU.*
- *Hypothesis 5b*: *COM positively influences PE.*
- *Hypothesis 5c*: *COM negatively influences PR.*

6.2.6 AES

AES dimension refers to the use of elements involved in clothing, such as the relationship between design principles and clothing. AES standard is an important criterion for consumers to evaluate clothing because clothing is an important means of visual communication (Liang, 2020). AES of clothing includes color, style, design, and other elements. AES attributes of clothing, such as color, style, and fabric, are the most important criteria for women to purchase clothing. In terms of smart clothing, Malmivaara (2009) believes that the AES factor is an important factor affecting the acceptability and wearability of the final product. Therefore, we propose that AES has a positive influence on ATT and PI:

- *Hypothesis 6a*: *AES positively influences purchase ATT of parent-child smart clothing.*
- *Hypothesis 6b*: *AES positively influences PI of parent-child smart clothing.*

6.2.7 ATT and PI

According to the theory of reasoned action, a person's action is determined by his intention to perform the action, and this PI is influenced by his ATT. In the context of clothing products, ATT has a positive influence on PI (Shaw, 2003). Therefore, we propose the hypothesis as follow:

- *Hypothesis 7a: Purchase ATT towards parent-child smart clothing positively influences PI.*

6.3 Method

6.3.1 Sample and procedure

In this study, an online questionnaire and offline interview were used to test the hypotheses and SEM with the obtained data. After eliminating the missing samples and incorrect data, a total of 372 samples was retained for research. Among this study, 74.14% of respondents were between the ages of 25 and 35, and 84.68% were parents of children. Among 372 respondents, 276 respondents said they had bought parent-child clothing (74.19%), and 254 respondents said they had bought smart wearable products (68.28%).

6.3.2 Instrument development

The measurement items in this research, based on previous studies, are developed and tested for reliability and validity. Eight potential components have been described in 28 measurement items, including FUN, COM, AES, PU, PE, PR, ATT, and PI. In this study, the variables, such as PU, PE, PR, FUN, COM, AES, ATT and others, are tested for reliability and factor analysis with the data obtained from the questionnaire and through SPSS software, and hypothetical model was drafted via Amos. Table 6-1 describes the survey items used in this study.

Table 6-1　Survey items in this study

Construct	Item	Measurement items
FUN	FUN1	The functions and features of smart clothing are stable
	FUN2	Using smart clothing is good to access contents and service
	FUN3	Interactive design functional clothing can enhance the fun interaction with children
	FUN4	Tracking and monitoring smart clothing can monitor the location of children in real-time to ensure their safety
COM	COM1	This product would be appropriate for my lifestyle
	COM2	This product can convey the family atmosphere
	COM3	This product would be more compatible with my current needs than the clothing I already have
	COM4	This product should focus on interaction with children and strengthen family integration

Continued

Construct	Item	Measurement items
AES	AES1	The choice of color for this product is very important
	AES2	The pattern design of this product is very important
	AES3	The overall design style of this product is very important
	AES4	The product looks professionally designed
PU	PU1	Wearing this product would improve the quality of my life
	PU2	This product promotes a better family atmosphere than ordinary family wear
	PU3	Wearing this product would increase my productivity
PE	PE1	The use of this product would improve the quality of my life
	PE2	Overall, find this product easy to use
	PE3	Using this product would not require a lot of mental effort
PR	PR1	How confident are you that the product/clothing will perform as described
	PR2	How certain are you that the product/clothing will work satisfactorily
	PR3	There are concerns about product safety
	PR4	Worry about the comfort of the product
ATT	ATT1	Smart clothes for kids is bad/good
	ATT2	Smart clothes for kids is unfavorable
	ATT3	Purchasing smart clothes for kids is foolish/wise
PI	PI1	I need to try this type of product
	PI2	It is likely that I will buy this product when it becomes available
	PI3	Using this product is worthwhile

6.4 Measurement model

6.4.1 Common factor extraction and naming

In the factor loading matrix, the absolute value of the factor loading indicates the degree of information overlap between the principal factor and the variable. The higher the degree of information overlap, the greater the generalized explanation ability of the principal factor is. Moreover, it is required that the common factor should be greater than 0.5. It can be seen from Table 6-2 that all the indicators meet the requirements. A total of 28 indicators can be classified into 8 categories, and all the measurement items are clearly classified.

Table 6-2 Correlations of the constructs and square root of AVE

Items	Items							
	FUN	COM	AES	PU	PE	PR	ATT	PI
FUN	**0.897**							
COM	0.379	**0.885**						
AES	0.084	0.083	**0.807**					
PU	0.558	0.401	0.091	**0.901**				
PE	0.541	0.257	0.11	0.509	**0.904**			
PR	−0.345	−0.27	0.012	−0.386	−0.353	**0.814**		
ATT	0.449	0.36	0.322	0.453	0.438	−0.328	**0.871**	
PI	0.465	0.366	0.129	0.465	0.489	−0.263	0.44	**0.87**

6.4.2 Measurement validity and reliability

The validity and reliability of measurement were analyzed by maximum likelihood estimation method for confirmatory factor analysis (CFA), full measurement model fits well, x^2/df is between 1 and 3, and the model has a simple adaptation degree. The root means square approximation error (RMSEA) is less than 0.08, and the model fits well. The three indexes of value-added fitness, TLI, IFI and CFI, are all > 0.9 indicates that the model fits well. The other two indexes of simple fitness, PGFI and PNFI, are all > 0.5 indicates an acceptable model. The estimated value extracted by load and average variance of each factor (AVE) reached the recommended threshold level of 0.60 and 0.50 respectively, which provided internal consistency and convergence. The square root of AVE of each construct was greater than the correlations between constructs, evidencing discriminant validity. Internal consistency for each construct was assessed using Cronbach's α. Cronbach's α coefficients for all eight constructs were acceptable, as they are all greater than 0.8. Results of CFA are summarized in Table 6-3.

Table 6-3 Factor loading and reliability of measurement Items and AVE

Constructs and Measurement	Standardized Factor Loading	Cronbach's α	AVE
FUN		0.943	0.8044
A4	0.916		
A3	0.882		
A2	0.907		
A1	0.882		

Continued

Constructs and Measurement	Standardized Factor Loading	Cronbach's α	AVE
COM		0.935	0.7829
B4	0.891		
B3	0.893		
B2	0.888		
B1	0.867		
AES		0.880	0.6514
C4	0.758		
C3	0.808		
C2	0.777		
C1	0.880		
PU		0.928	0.8112
D1	0.907		
D2	0.899		
D3	0.896		
PE		0.930	0.8165
F1	0.922		
F2	0.914		
F3	0.874		
PR		0.881	0.6633
G1	0.739		
G2	0.924		
G3	0.854		
G4	0.724		
ATT		0.912	0.7587
H1	0.822		
H2	0.903		
H3	0.886		
PI		0.909	0.7560
Y1	0.893		
Y2	0.898		
Y3	0.815		

Chapter 6
Case Study: Purchase Intention and Design Elements of Parent-child Clothing

6.5 Hypotheses testing and inspiration

Figure 6-2 shows the structural equation model results obtained by confirmatory factor analysis of variance with Amos, which reflects the causal relationship between various latent variables. The model in this paper contains eight factors (latent variables): FUN, COM, AES, PU, PE, PR, ATT, and PI.

As can be seen from Table 6-4, 12 of the 14 regression coefficients directly affected are significant.

As can be seen from Figure 6-2 and Table 6-4, in the independent variable, the P value of COM to PE is less than 0.05, reaching the significance level of 0.05, and the coefficient is

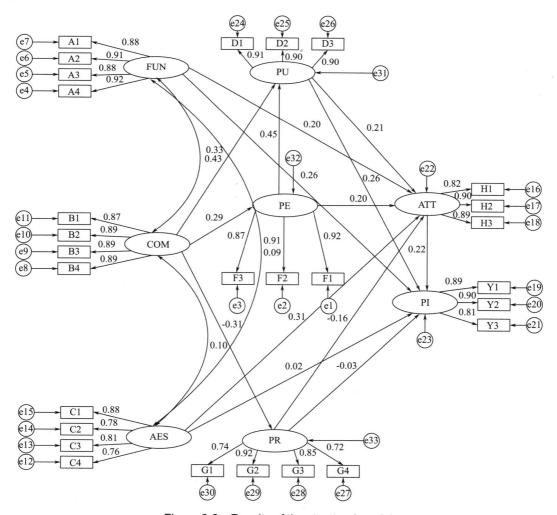

Figure 6-2　Results of the structural model

positive, indicating that COM has a significant positive influence on PE, so Hypothesis 5b is established.

The *P* value of COM and PE to PU is less than 0.05, reaching the significance level of 0.05, and the coefficient is positive, indicating that COM and PE have a significant positive influence on PU, so Hypotheses 5a and 2a are established.

The *P* value of COM to PR is less than 0.05, reaching the significance level of 0.05, and the coefficient is negative, indicating that COM has a significant negative influence on PR, so Hypothesis 5c is supported.

The *P* value of FUN, AES, PU and PE to ATT is less than 0.05, reaching the significance level of 0.05, and the coefficient is positive, indicating that FUN, AES, PU and PE have a significant positive influence on ATT. So, Hypotheses 4a, 6a, 1a, and 2b are established.

The *P* value of PR to ATT is less than 0.05, reaching the significance level of 0.05, and the coefficient is negative, indicating that PR has a significant negative influence on ATT. So, Hypotheses 3b is established.

Table 6-4 Standardized latent variable path coefficient

Path			Estimate	S.E.	C.R.	P
PE	<---	COM	0.294	0.063	5.436	***
PU	<---	PE	0.446	0.045	9.031	***
PU	<---	COM	0.325	0.051	6.675	***
PR	<---	COM	−0.315	0.055	−5.574	***
ATT	<---	FUN	0.203	0.043	4.159	***
ATT	<---	AES	0.306	0.059	5.982	***
ATT	<---	PU	0.211	0.052	3.572	***
ATT	<---	PE	0.196	0.046	3.399	***
ATT	<---	PR	−0.162	0.047	−3.305	***
PI	<---	FUN	0.260	0.054	4.997	***
PI	<---	AES	0.017	0.072	0.322	0.748
PI	<---	PU	0.255	0.057	4.657	***
PI	<---	ATT	0.219	0.074	3.461	***
PI	<---	PR	−0.031	0.057	−0.613	0.540

Note: *** $P < 0.001$.

The *P* value of FUN, PU and ATT to PI is less than 0.05, reaching the significance level of 0.05, and the coefficient is positive, indicating that FUN, PU and ATT have a significant positive influence on PI. So, Hypotheses 4b, 1b, and 7a supported.

The *P* value of AES and PR to PI is larger than 0.05, failing to reach the significance level of 0.05, indicating that AES and PR have no significant influence on PI. Then Hypotheses 6b and 3b are not established.

6.6 Summary

According to the above hypotheses' verification, both FUN and PU have a positive effect on purchase ATT and PI, because the consumers would like to purchase the product when it's FUN meets the needs of the consumers.

AES positively influences purchase ATT but doesn't work well in PI, it may be that for products such as parent-child smart clothing, AES is one of the factors influencing consumers' purchase desire, but it will not directly lead to purchase behavior. Because other ordinary parent-child clothing can also meet AES characteristics, the parent-child smart clothing is not very dominant in emphasizing AES (Perry, 2017; Chattaraman, 2006).

PR negatively influences purchase ATT, however, the negative influence of PI is not significant. We believe that the technical quality, washing and care, sensing and contact of parent-child smart clothing are all purchase factors that consumers consider (Chan, 2012), but they do not have a great influence on purchase decision. Its innovation, technology, and other characteristics, which are different from ordinary parent-child clothing, mainly affect PI.

COM positively influences PU and PE, and negatively influences PR. PE positively influences PU. COM includes innovation, which indicates that, when consumers think that they will have the desire to buy if the product is easy to operate, well used by children and meet the demands of the consumers (Wright, 2014). It may also mean that because of the simplicity of the operation, it will not cost more time for consumers to learn how to use new technologies. Consumers who have bought similar products in the past will be less risky if they have some experience using them.

In this study, in addition to a large number of online questionnaires, face-to-face interviews were conducted offline to further understand consumers' views on parent-child smart clothing. The participants included three college fashion design teachers, four staff supervisors of children's clothing design companies, and three information engineering programmers. It is learned that all of the 10 participants have the experience of buying smart products and parent-child outfits. We invited participants to the clothing Lab of the university to conduct the experiment by e-mail. Based on the results of the online questionnaire survey, we ask the participants to further discuss the hypothetical factors such as interest, aesthetics, and risk, and put forward some design factors that may affect consumers' purchase. The overlapping answers discussed by the participants were listed and scored with the Likert scale. The full score is 5 points, which are: very satisfied 5 points, satisfactory 4 points, general 3 points, dissatisfied 2 points, and very dissatisfied 1 point.

Table 18 shows the scores of 10 participants and online volunteers.

Based on the results of previous online questionnaires, a discussion was conducted around the hypothetical factors such as FUN, AES and PR are discussed, and some design factors that may affect consumer purchase are put forward and scored by Likert scale. The full score is 5 points, which are: very satisfied: 5 points, satisfied: 4 points, general: 3 points, dissatisfied: 2 points and very dissatisfied: 1 point respectively. Table 6-5 shows the results scored by 10 participants and online volunteers.

Table 6-5　Design element rating

Functionality	The average	Aesthetic	The average	Perceived performance risk	The average
Interaction design	4.2	The color	4.05	Worry about wash protect	4.19
Tracking	4.14	The fabric	3.99	Worry about comfort	4.2
Sports health	4.18	Version	4.2	Worry about the quality	4.05
Comfort	4.42	Style	4.01	Sensing the contact	4.03

All hypotheses are established except 3b PR and 6B AES. Although the effects of AES and PR of PI are not significant, they have a positive influence on consumers' ATT. Therefore, it is understood that compared with ordinary parent-child clothing, the technical level of parent-child smart clothing is also an important factor affecting consumers' purchase besides the ordinary external attributes of clothing. Accordingly, it is required that parent-child smart clothing should balance design and technology to make it comfortable and fashionable and improve practicality.

We find that parent-child smart clothing can be divided into two levels: technical attribute and clothing attribute. Combining with the results of Table 6-4 and Table 6-5 and relevant suggestions provided by the participants, we can deduce the framework model of design elements (Figure 6-3). Among them, there are three elements of clothing attribute (the left side of Figure 6-3): COM, FUN, and AES. It is suggested that when designing parent-child smart clothing, researchers should consider whether it can meet the needs of children and parents, conform to their lifestyle, representativeness and transmissibility in terms of COM. In terms of FUN, interaction design, monitoring and tracking, exercise and health can be considered. As can be seen from Table 6-5, for parent-child smart clothing, consumers hope that practical and comfortable clothing can be worn at the same time. Among them, the average score of interaction design is 4.2, which is a favorable function for consumers and can stimulate the mutual affection of wearers (Wang, 2018), so designers can consider to make more use of this function in design. Although AES has little influence, which should not be ignored by the designers. As can be seen from Table 6-5, the type of clothing will be the most important factor for consumers, followed by colorless. Therefore, we suggest that when designing parent-child smart clothing, the designer should pay more attention

Chapter 6

Case Study: Purchase Intention and Design Elements of Parent-child Clothing

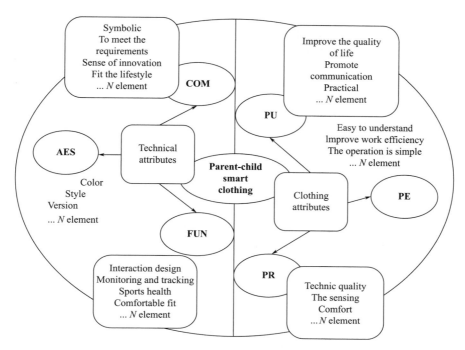

Figure 6-3   Framework model of design elements

to the control of the version and color of the product in addition to insisting on own style, and differing from the single and boring parent-child clothing in the market.

There are three elements of the technical attribute (the right side of Figure 6-3): PU, PR, and PE. Different from ordinary smart clothing, parent-child smart clothing is used by parents and children together, which should consider whether the children can master the use of the product. Therefore, the designer should consider whether the operating system of the product is complex, and try to make it easy to understand, so that young children can use it skillfully and easily.

Researchers should not only work hard on the external attributes of clothing, but also make sure that the products can improve the life quality, facilitate communication between children and parents, and operate in a simple and easy way, with high quality. Additionally, common problems such as washing and care, sensing and touching of smart clothing should be seriously considered. Although PR has little influence on PI, it also affects consumers' ATT. Therefore, this factor shall not be ignored by the designers.

Through Hypothesis 3b, it is learned that the technical quality, washing and care, sensing and touch of parent-child smart clothing and other issues are all factors that consumers consider to buy, but they have little influence on the decision to buy. Its innovation, science and technology and other factors that are different with ordinary parent-child clothing are the factors that affect the consumers may buy. Therefore, the designers should pay attention to the injection of technology elements to make clothing innovation more interesting.

We understand and find that consumers' concerns about parent-child smart clothing and the comfort children wear. The body quality of the children at low age is weaker and the skin is tender. Thus, the comfort of children's clothes is a factor that parents should take into consideration before purchasing. Accordingly, the designers should add technology elements while maintaining the comfort of clothes. Since PU is also a factor affecting consumers' PI and attitude, the designers should ensure that products are practical and meet consumers' needs. At present, communication and interaction are what parents and children need most, so the designers can consider more interesting interaction design elements into the design. In general, the researchers should not only work hard on the external attributes of clothing, but also make sure that the products can improve the life quality, facilitate communication between children and parents, and operate in a simple and easy way, with high quality. Quality should also be guaranteed, common problems of smart clothing, such as washing and care, sensing and contact should be seriously considered.

To sum up, designers who want to be involved in parent-child smart clothing, both technical attributes and clothing attributes need to be taken into account. When we generally pay attention to smart clothing in areas such as intelligent medical and health care, very few people will be involved in interesting smart parent-child costumes. For children, the way they communicate and get along with their parents is very important. Increase the interaction time between children and parents through interactive smart parent-child clothing, and easily shorten the psychological distance between children and parents in an interesting technological way. Attempts may be made to adopt simple but interesting means of interaction for this interactive parent-child clothing such as gesture sensing, distance sensing and voice control. For example, the colored lights on the clothes will gradually light up when the father and the child come into contact with each other. Or when the mother approaches the child, the child's favorite music will play. Surely, in addition to simple functionality, other factors which affect consumers' purchase decision should be also taken into account i. e. comfortableness of fabric, weight of sensor, safety and whether it's designed for daily use in terms of style.

Chapter 7

Knowledge Transfer and Enlightenment to Interactive Clothing Design

7.1 Knowledge transfer

The knowledge structure of interactive clothing has been preliminarily formed up to this point. According to the principle of Socialization-Externalization-Combination-Internalization (SECI) model (Nonaka & Takeuchi, 1995/1996) which has become the cornerstone of knowledge creation and transfer theory, in the first four stages of this study, the tacit knowledge and explicit knowledge related to interactive clothing were explored respectively, and the related knowledge system was formed preliminarily.

As shown in Figure 7-1, this research adopts the five-phase SECI model of the knowledge-creation process to architecture the knowledge structure in five phases: ①sharing tacit knowledge; ②creating concepts; ③justifying concepts by prototyping; ④evaluating prototypes; ⑤cross-leveling knowledge.

- **Phase 1: Sharing tacit knowledge**

    In the stage of literature theory analysis, i. e., the socialization of knowledge, this research obtains tacit knowledge 1, i. e., the humanistic attributes of interactive clothing, through the integration of needs psychology, clothing sociology and clothing design hierarchy analysis. By comparing the IoT data evolution theory, CPS architecture and interactive clothing development technology, tacit knowledge 2, i. e., the technical attributes of interactive clothing, is acquired.

- **Phase 2: Creating concepts**

    Corresponds to externalization, the conceptual model of the technological process of interactive clothing is summarized based on the tacit knowledge 1 and 2 acquired in Phase 1, and the explicit knowledge 1, namely CPCS model, is acquired.

- **Phase 3, 4: Justifying concepts by prototyping and evaluating**

    New concept created in the previous step needs to be justified at some point in the procedure.

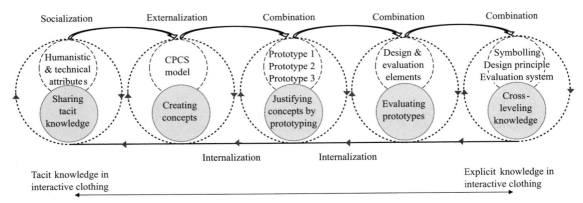

Figure 7-1 Knowledge transfer of this research (after Nonaka & Takeuchi, 1995)

Three types of clothing prototypes, including smart infant clothing, interactive couple clothing, and parent-child clothing, successfully validated the feasibility of explicit knowledge 1, i.e., the CPCS model, and formed a new explicit knowledge 2, i.e., the prototype design and development methods. Through the evaluation of the prototype, re-mining tacit knowledge 3, i.e., the design optimization elements and evaluation elements.

• **Phase 5: Cross-leveling knowledge**

This is the process by which this study transforms explicit concepts into knowledge systems during the combination of knowledge. Three essential explicit knowledge including interactive clothing signal and symbol conversion, design principles, and design evaluation framework. As a result, the tacit knowledge gradually transforms to form the explicit knowledge of interactive clothing. And this explicit knowledge may trigger a new round of interactive clothing innovation or the exploration of tacit knowledge responses to internalization.

7.2 Completion of loops steps in the CPCS model

The process of above humanistic evaluation is also an extensive process of knowledge mining in the CPCS model. The conclusion of humanistic evaluation is the cognition of interactive clothing design optimization, which is also the summary of wisdom. Wisdom further guiding the optimized setting and measurement of configuration basis signals, thus forming the complete nine loops steps of interactive clothing CPS. These are responses to the SQ2 presented in the introduction section.

The experiment progressed to this step, where the technical process of the entire CPCS architecture with parent-child clothing as a case was fully implemented.

• **Potential applications of such models**

In the development process of the interactive clothing, step ⑨ in Figure 3-2 is to perform the following two steps according to the evaluation results of the previous step. The first is to improve the social symbolic performance of clothing in interpersonal interactions, for example, the diversified message transmitted by special changes such as styles and colors exhibited by clothing can generate more social communication symbols (Devendorf et al., 2016; Kan et al., 2015). Secondly, the microprocessor embedded at the physical level is configured, an improved task to guide the optimized form of the signal setting, such as the mining of the brain wave or other human body's physiological and even psychological signals (Kosti et al., 2018).

- **Summarize the symbolization of the signal**

The experimental results demonstrate that the CPCS model has a universal guiding significance for various types of interactive clothing development, which solves the problem that the CPS architecture of interactive clothing should have wide applicability mentioned above in Chapter 1. In addition, the CPCS model takes the correlation analysis of CPS architecture and DIKW level as the icebreaking and provides systematic research and development roadmap for the development of interactive clothing, which also can response to the existing dilemma mentioned above in Chapter 1.

For the SRQ 2 of this research to be answered, the development process from information to knowledge is the evolution from the data of information science to the wisdom of humanistic science, that is, the process of symbolization is widely recognized by social groups. As shown in Figure 7-2, the four components split into information science behaviour (data, signal) for technical conversion steps and the humanistic science behaviour (wisdom, symbol) for humanistic cognition steps, they constitute the basic structure of information transformation into knowledge.

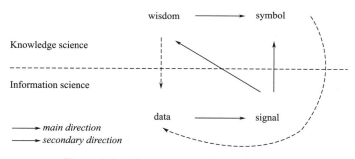

Figure 7-2 The symbolization of the signal

- When the data extracted from the clothing or the wearer's body are converted into a digital signal, it can be transformed into a wisdom content that is accepted by people, such as the experience, and this experience will also feedback and change the developer's settings for the data probe of clothing.

- The wisdom of the social dimension will eventually develop into a widely recognized symbol of clothing sociology or clothing psychology.
- The signal can also be directly converted into a social symbolic-message (Jeon, 2017) by scheduling the big data algorithm.
- The social symbols may have a feedback impact on data measure standards.

The starting point of the whole process is the physical data input of clothing or the physiological data input of the wearer, and the end point is the symbolic output of clothing sociology or psychology, that is, the symbolization of clothing-mediated signals.

• **Summarized the process from signal input to symbol output**

Combined with the parent-child clothing development process and the CPCS model, it can be found that there are 9 significant steps from the input of the physical signal to the output of the social symbol (Figure 7-3): 1, data; 2, signal; 3, conversion; 4, machine learning (A-computing, B-communication, and C-control); 5, data computation (A'-time, B'-task, and C'-scheduling); 6, knowledge mining; 7, humanistic evaluation; 8, interaction; and 9, symbol. Among them, signal input, humanistic evaluation, and symbol output steps are the most critical hubs in the process. From 1 to 6 steps belong to the field of information science, and 7 to 9 steps belong to the field of knowledge science.

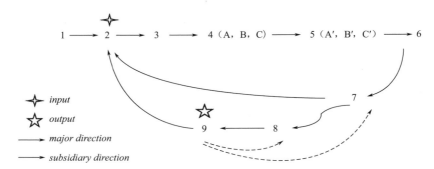

Figure 7-3 The process from signal to symbol

In the process of bridging the connection between the signals in the field of information physics and the symbols in social dimensions, the machine learning (computing, communication, control) and data computation (time, task, scheduling) undertook the information physics technical surgery, while humanistic evaluation, interpersonal interaction, and social symbol output undertook the social dimension of the task. In the stage of knowledge formation, the evaluation object should be optimized according to the feedback message of social interaction symbols, to promote the redesign of physical signals to optimize the interpersonal interaction effects of interactive clothing. These elements once again responses to the SRQ 2 mentioned in Chapter 1 of this research.

7.3 The "C" design principle for interactive clothing

Academic research on the design process and development elements of smart clothing is still inadequate. In the limited literature, relatively complete research focuses on the model summarized by Lee (2016), that is, Cradle-to-Cradle Design Framework (Figure 7-4) which is an integrated framework focused on sustainable design and evaluation processes of smart clothing design. This model integrates many theories respectively: McDonough & Braungart (2002) introduced the Cradle-to-Cradle Design Model, Lamb & Kallal (1992) proposed the Functional-Expressive-Aesthetic Consumer Needs Model, Kelly (2016) described the WEarable Acceptability Range Scale, and Ajzen (1991) revealed the Theory of Planned Behavior. However, it can be found from the architecture level that this model only focuses on the human perspective of the target users in the design process from attitudes toward the behavior, and there are no specific indicators in the

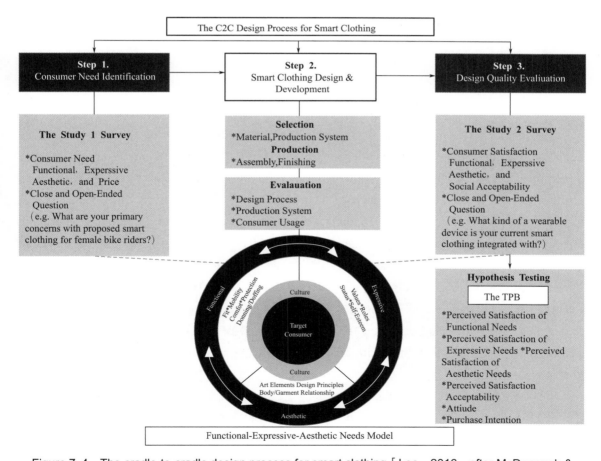

Figure 7-4　The cradle-to-cradle design process for smart clothing [Lee, 2016, after McDonough & Braungart (2002), Lamb & Kallal (1992), Kelly (2016) and Ajzen (1991)]

application level of information technology in fashion design. Therefore, the model cannot be called a design system, and there is no complete structure level and design index. So, what are the design criteria or principles of interactive clothing?

By summarizing the implementation process of the prototype experiments in case study 2 and 3, 18 design principles of interactive clothing with letter "C" as the first letter can be distilled. We divided the interactive clothing design architectures into three factors as levels, links and contents are shown in Table 7-1. These elements response to the SRQ 3 mentioned in Chapter 1 of this book.

Table 7-1 The "C" principles and architecture of interactive clothing design

No.	Levels	Links	Contents
1	Essential	Subject aspects	(1) Clothing; (2) Consumer-centered; and (3) Culture symbol
2	Technical	Information engineering aspects	(4) CPCS; (5) Connection; (6) Conversion; (7) Computing; (8) Communication; (9) Control; (10) Cognition; and (11) Configuration
3	Humanistic	Art design aspects	(12) Critical thinking; (13) Creativity
4	Objective	Realization form aspects	(14) Communication; (15) Convenience; (16) Characterize; and (17) Captivating symbol
5	Organizational	Personnel aspects	(18) Collaboration

The logical relationship between them is shown in Figure 7-5. With the dressers and the potential cultural semantic representation of clothing (in the center of Figure 7-5) as the core, the art design team of interactive clothing (on the right side of Figure 7-5) and the information engineering development team (on the left side of Figure 7-5) are respectively formed and engaged in interactive collaboration. As a new thing, the art design of interactive clothing should adopt critical thinking to get creative, and the realization of its interaction effect cannot be separated from the application and development of CPS technology in the field of information engineering. The purpose of the final realization of interactive clothing, the effect of rendering, the distinctive characteristics and attractiveness to the user (Figure 7-5, peripherals), is to achieve the new form of communication with clothing as the medium and to provide convenience for life, to enrich its characteristics, and to attract the desire to acquire interactive clothing. The meaning of each "C" principle is as follows:

(1) **Consumer-Centered.**

The target consumer (intended wearer) is at the core of the model. The development of interactive clothing should be human-centered and take people's social communication as the foundation. Clothing should serve the communication needs of people, not the people to serve to clothe.

Chapter 7
Knowledge Transfer and Enlightenment to Interactive Clothing Design

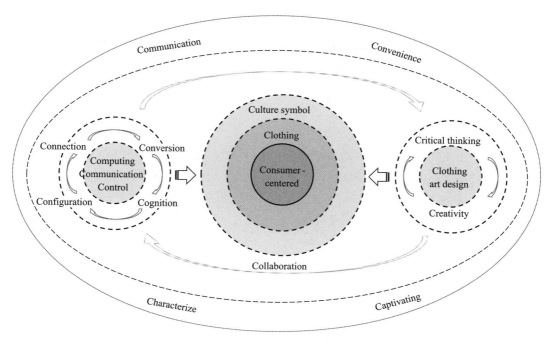

Figure 7-5 The logical relationship of "C" principle for interactive clothing design

Interactive clothing should be able to meet the wearers' psychological needs of Esteem and Self-Actualization level. It should be consistent with the meaning of the high level in the model of Figure 2-2.

(2) **Clothing.**

The essence of interactive clothing is still clothing. It is a kind of clothing with a technical function attached, not a mere smart device, let alone a machine. Advanced interactive clothing should be consistent with the meaning of the high level in the model of Figure 2-12.

(3) **Culture symbol semantic.**

In the symbolic semantic expression of clothing, interactive clothing should meet the standards of social functions such as conveying certain social and cultural connotations and facilitating communication and expression. It needs to be aligned with the meaning of the high level in the model of Figure 2-2.

(4) **Critical thinking.**

For the design of innovative interactive clothing that upsets the concept of traditional clothing, its inspiration needs critical thinking. Regarding design categories to reach the level of designer custom-made, while also emphasizing creativity, originality and high added value, embodying the characteristics of pop and fashion to provide customized design services for niche customers. It should be consistent with the meaning of the high level in the model of Figure 2-4.

(5) **Communication.**

The function of interactive clothing is to realize the two-way interaction between the human body and clothing, wearers and their environment. Move closer to the level of the IoC concerning functionality. It needs to be aligned with the meaning of the high level in the model of Figure 2-7.

(6) **Convenience.**

The purpose of interactive clothing is to provide convenient services for the interpersonal communication of the dressing person in daily life. Converge to service in DIKWS to achieve that data as a service, information as a service, knowledge as a service, and wisdom as a service on the whole (Figure 2-9).

(7) **CPCS.**

Under the CPCS framework, the technical means of information processing optimize the computing, communication and control algorithms, to realize the data cycle processes from the connection, conversion to cognition and configuration (Figure 3-2).

(8) **Collaboration.**

Cross-domain team-working research and development. During the three prototyping process, the development team members were drawn from the fields of Information Engineering, Optoelectronic Science, Garment Engineering and Fashion Design respectively. According to the results of those prototype development tests, it is also necessary to introduce a professional team of Textile Engineering to realize the research and development of smart textiles.

(9) **Characterize.**

Embodying the characteristics of high-level bidirectional interaction. Two interactive clothing case prototypes only use distance sensing to drive the appearance of clothing changes in the form, and increasingly iterative artificial intelligence technology can enrich the form of clothing interaction. In the aspect of signal input, the existing mature technology such as blood pressure pulse sensing, brain wave sensing and other technical forms can be applied to interactive clothing development. Regarding symbol output, the symbolic semantic of interactive communication can be diversified to form richer features. It should be consistent with the meaning of the model Figure 6-2.

(10) **Creativity.**

Creative design and aesthetic value. The development of smart clothing has completed the technology-centered of the most basic technology reserve stage. Interactive clothing development should be divorced from the limitations of engineering technology development and enhance creativity from the perspective of "taking Art and Humanities as the center."

(11) **Captivating.**

The ultimate effect of clothing should be attractive and can arouse the desire to own it.

7.4 The framework of the evaluation criteria for interactive clothing design

7.4.1 Background and significance

The evaluation system refers to an organic whole internal structure, which is composed of multiple indicators representing the characteristics of evaluation objects in various criteria and interrelated elements. Up to now, although some scholars put forward individual evaluation methods for case studies, the academic evaluation system of smart clothing has not been established, and interactive clothing as a new type of smart clothing cannot expect the existence of targeted evaluation system.

This research has summarized in the CPCS model that the current design evaluation direction for smart clothing or interactive clothing are mainly divided into three dimensions, that is effective indicators (emotion, five senses, etc.), ergonomics indicators (action, posture, comfortable, etc.), and functional indicators (sport monitoring, healthcare, entertainment, etc.). But in the specific case study, this research only adopts the evaluation method of Kansei Engineering, and cannot systematically evaluate the effect of the prototype.

Based on the above academic situation and summary of the case implementation experience in this research, in order to fill the academic gap, this research proposes to build the basic structure of interactive clothing design evaluation system of the framework level.

In addition, the purpose of establishing the hierarchical structure of the evaluation system is not merely to evaluate the ranking and advantages and disadvantages of clothing works, but also to guide the improvement of the evaluation system of interactive clothing in the future, and to encourage the object of evaluation to develop in the right direction and objectives.

7.4.2 The structure of the evaluation criteria

7.4.2.1 Theoretical basis and method

When designing the evaluation criteria system, we should first have the scientific theory as the guidance. So that the evaluation criteria system can be rigorous and reasonable in the basic concept and logical structure, grasp the essence of the evaluation object and be targeted.

According to "Evolutionary route based on psychology, sociology and design" (Figure 2-14) and "Evolutionary route based on IoT and CPS applications" (Figure 2-15), the two models summarized the characteristics and attributes of interactive clothing in psychology, sociology and design from the humanistic point of view, and summarized various characteristics and attributes of IoT and CPS applications from the technical point of view. Moreover, the evolution and development of interactive clothing should not be lower than the fourth layer attribute or function in the model, and we use these characteristics and attributes as the main reference standard for the design and evaluation architecture of interactive clothing.

The method of designing the evaluation criteria system adopts the Analytic Hierarchy Process (AHP), which is decomposed into sub-indicators by the total index and then decomposed into sub-indicators by the sub-index. Usually, people call these three levels as the target level, the criterion level, and the indicator level. As a reference, we divide the hierarchy of interactive clothing evaluation system into five levels: Target level, Perspective level, Criterion level, Index level, and Element level.

7.4.2.2 The frame structure of the evaluation criteria

According to the AHP method, the frame structure can be divided five levels according to the following ideas (Table 7-2). These elements also response to the SRQ 3 mentioned in Chapter 1 of this book. The relationship between this five-layer structures can be found in model Figure 7-6.

Table 7-2 AHP of the interactive clothing evaluation system

Level	Purpose	Meaning	Content
Level 1	Evaluation objectives	Target level	Interactive clothing
Level 2	Evaluation perspective	Perspective level	Humanistic & technical
Level 3	Evaluation form	Criterion level	Subjective & objective evaluation
Level 4	Evaluation guidelines	Index level	Emotion & aesthetics; function, ergonomics & emerging technologies
Level 5	Evaluation factor	Element level	Cognition, perception, associational, Communication, Esteem, Self-actualization, Expressive, Interactivity, Communicative, Auxiliary, Comfort, Temperature, Pressure, Action Posture; Textile Science, Sensing, communication, response control, Deep learning, ⋯ N element

- Level 1 structure: evaluation objectives, that is, interactive clothing.
- Level 2 structure: evaluation perspective. Take interactive clothing as the main body, with humanistic and technical perspective as the wings.
- Level 3 structure: evaluation form and criteria. Subjective evaluation and objective evaluation.
- Level 4 structure: evaluation guidelines and index. Subjective indexes include emotion and aesthetics, and the objective index includes function, ergonomics and the application of emerging technologies, etc.
- Level 5 structure: evaluation factors. Among them, the Emotion index includes expressiveness, desire to own, communication, esteem, self-actualization and other elements.

Aesthetics index includes cognition (formal), perception (sensory), associational (meaning and values), etc. Among them, cognition (formal) factor mainly for clothing styles and colours, which in turn involves the formal beauty of clothing design factors, such as contrast, symmetry, scale, rhythm, and emphasis elements.

Function index includes interactivity, communication, auxiliary and other elements.

Ergonomics index includes Comfort, temperature, pressure, action, posture, and other elements.

The new technology index, including textile science, sensory, communication, response control, deep learning, and other elements.

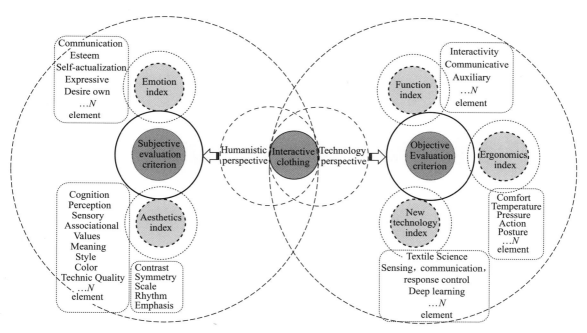

Figure 7-6 The structural relationship of evaluation criteria for interactive clothing design

Chapter 8
Conclusion

The main differences between the contents of this book and previous studies by other scholars are as follows:①this book pioneered a new field of smart clothing research, namely interactive clothing; ②this book introduces CPS technology into clothing development from the perspective of art design; ③a research framework for interactive clothing was initially established.

The interactive clothing, a topic on which there is little previous research, an evolutionary branch of smart clothing in the field of information science, which emphasizes the function of social symbols that mutual interaction or communication between the wearer and their environment based on the integration of information science and traditional clothing. Combining CPCS with clothing engineering design to input a certain physical signal into clothing, the interactive clothing can output a specific social symbol that people or clothing environment can perceive and generate corresponding interaction. This kind of social symbol is the expression form of the interactive clothing as the interactive medium for people to interact with the environment.

With the gradual integration of IoT/CPS into people's daily life and the growing development of smart textile technology, the revolutionary evolution has been occurring at an unprecedented rate in the fields of clothing; interactive clothing will play an increasingly important role in future interpersonal communication and interaction.

The research on interactive clothing should integrate the two opposing perspectives of humanities and technology, and bridge the gap between humanities and technology from an interdisciplinary perspective in the process of developing and evaluating the prototype of interactive clothing. Because in essence, interactive clothing is still clothing, and the interactivity is an additional feature. No matter how complex and changeable the interaction is in technology, it should not be divorced from the material and social attributes of clothing as well as the material and spiritual functions of clothing.

8.1 Contribution

8.1.1 Main practical contribution of this book

8.1.1.1 The contribution of prototyping practice

An overview of the whole practice process of this book, the implementation of the prototype experiment began in 2012 and achieved progressive results in three stages. These prototypes were developed from easy-to-difficult design processes and interdisciplinary research methods provide a fruitful practical application reference for designers who are engaged in the art design field but not familiar with the relevant information technology.

- The first stage was in 2013, two types of infant's smart clothing were developed. Analyze the feasibility of the application of IoT technology in the design and development of garment products from the perspective of sensor technology application. The purpose of the two prototyping experiments is to combine IoT sensing technology with infant clothing design, to develop a real-time monitoring alarm clothing for infants. As a result, two of China's utility model patents have been granted, which illustrates the feasibility and rationality of the original planning of this research. Furthermore, the fundamental study of interactive clothing has established the technical foundation by these experiments.
- The second stage is which interactive clothing for couples has been made from 2016 to 2017. It is a preliminary study for IoC. The aim is to bridge the gap between human emotions and wearable technologies for interactive fashion innovation, to consider the reasons why smart clothing should satisfy the IoT technical functions and human emotional expression simultaneously, to investigate the manner in which artistic design perspectives and engineering methods combined effectively.
- The prototype development of interactive parent-child clothing is the third stage from 2017. The aim is to bridge the gap between CPCS architecture model and emotional evaluation, to explain how the transformation could be realized from information to common feeling, to refine the design elements for interactive clothing.
- At the stage of the following two prototyping experiments, in the aspect of practical research, we focus on the existing problems and research motives mentioned in Chapter 1, two series of interactive clothing prototypes were successfully designed and produced, and Kansei evaluation experiments and data analysis were carried out. And the theoretical research focuses on the research aims and objectives mentioned in section 1.3 of Chapter 1, defines the definition of interactive clothing, and build many innovative theoretical models that can verify each other with the prototype from the perspective of humanities and technology as well as their integration.

8.1.1.2 Contribution to the clothing industry and the aging society

This topic expands the industry chain of the clothing industry, not only integrating the main elements of traditional clothing industry but also adding CPS/IoT technology, which makes clothing and clothing, clothing and environment, clothing and people interact. More importantly, it puts forward the concept of the internet of clothes (IoC), that is, the industrial prospect of interaction between people in social groups through clothing medium.

At the application level, this topic focuses on the social needs of the next 10-20 years, especially the interactive clothing needs of the aging society are of great significance. With the continuous development of the information society, Smart Home, Smart City, and intelligent transportation will bring more convenience to our lives. At the same time, in the face of the elderly's living needs, emotional needs, communication needs, interactive clothing will become a vital carrier medium in the future.

8.1.2 Original contribution to knowledge science

8.1.2.1 Innovative concept of interactive clothing and IoC

This book pioneered the concept and definition of interactive clothing in academia, puts forward its origin and function, explains its technical characteristics in the view of information science and the social characteristics from the perspective of the humanities science through the R&D roadmap model and development practice of a series of prototypes. Also, explored the development direction of smart clothing in the future and analyzed the possibility of realizing IoC based on CPS/IoT technology.

8.1.2.2 Models for researching the humanistic and technical attributes of interactive clothing

At the psychological level of humanistic perspective, interactive clothing needs to meet the "Self-esteem needs", and it is necessary to satisfy the function of "Connotation" in the sociology of clothing, and it belongs to the "Couture fashion" category in clothing design typology.

From the technical perspective, interactive clothing needs to meet "Wisdom" requirements in IoT data theory, "Cognition" functionality in the 5C architecture of CPS, and the transmission of "Humanistic symbols" regarding technical processes.

8.1.2.3 Constructing an innovative Cyber-Physical-Clothing Systems architecture

At present, the academic has not yet proposed a complete technical process model for the prototype development of smart clothing or interactive clothing. This research pioneer summarized the flow of interactive clothing design and development and revealed the technical relevance of interactive clothing, DIKWS, and CPS in the CPCS model. And also, this research verified its feasibility with the interactive clothing prototyping as a case.

The role of this CPCS model is to explain the technical development process of interactive clothing from clothing to data, information, knowledge, wisdom, services, humans, and then back to clothing (C2H for short) in the perspective of CPS and DIKWS architecture. The interactive clothing has the characteristics of information processing and human feedback, it is controlled by machine-based algorithms and integrated the network with its wearers into one unified holistic framework, which determines its interactive architecture can be divided into three levels: physical level, cyber level, and social level. Each level uses different techniques and has different effects. This model provides specific technical guidance for the development of interactive clothing and even prototypes involving larger categories of smart clothing.

8.1.2.4 Revealed the innovative "C" design principle

Revealed the design principles of interactive clothing from the perspective of humanities and technology respectively, and guided the design of interactive clothing.

By summarizing the implementation process of case studies, 18 design principles of interactive clothing with letter C as the first letter was distilled. The specific links include essential aspects, information engineering aspects, art design aspects, objective aspects, and organizational aspects. The specific elements include:①Clothing; ②Consumer-centered; and③Culture symbol; ④CPCS; ⑤Connection; ⑥Conversion; ⑦Computing; ⑧Communication; ⑨Control; ⑩Cognition; ⑪Configuration; ⑫Critical thinking; ⑬Creativity; ⑭Communication; ⑮Convenience; ⑯Characterize; ⑰Captivating symbol; and⑱Collaboration.

8.1.2.5 Established a framework of evaluation criteria for interactive clothing design

Established a basic framework of interactive clothing design evaluation criteria system, to encourage the object of evaluation to develop in the right direction and objectives. Up to now, although some scholars have put forward individual evaluation methods for case studies, the academic evaluation criteria of smart clothing have not been established, and interactive clothing as a new type of smart clothing cannot expect the existence of targeted evaluation criteria.

We divide the hierarchy of interactive clothing evaluation criteria into five levels: Target level, Perspective level, Criterion level, Index level, and Element level. This framework takes interactive clothing as the main body, with humanistic and technical perspective as the wings. Subjective evaluation criteria consist of emotion and aesthetics indexes, and objective criteria include function, ergonomics and the application of emerging technologies indexes, etc. Also, each indicator contains multiple subdivided elements.

8.2 Future research

Although the research of this project has achieved several results in both theoretical and practical links, there are still limitations. To be ready implemented in many fields of interpersonal interaction realized through clothing media, some extended research is crucial to be done, and future research needs to be focused on the following issues:

- It needs to continue to develop the automatic interactive response function of clothing made to the wearer's physiology and psychology.

In the prototype development process of this project, case prototype 1 used temperature and humidity sensing technology, case prototype 2 and 3 only used distance sensing. The interaction effects of case prototype 2 and 3 are limited to clothing, not interactions between people. This sensing technology cannot reflect the physiological changes of the wearer, nor can it reflect the interactive communication caused by the physiological and psychological changes of different wearers. As a result, emerging technologies such as electrocardiography (ECG) need to be introduced into prototype development.

This book use a questionnaire for emotional evaluation of Kansei methods, but it is basically same with SD methods. The technology of affective computing, including recognition of facial expression or bio-sensor for a rousing, will be adopted for the CPS & IoT application research in future.

- It needs to continue to conduct interdisciplinary research, establish a database of human physiological and psychological responses related to clothing life, and continue to optimize the CPS algorithm.

Similar to autopilot technology, it can automatically execute different procedures based on different environmental changes and safely arrive at the destination. In order to realize that the individual or group of dressers can be reflected instantly through the clothing medium according to different physiological and psychological reactions in the process of interpersonal communication, it is necessary to establish a huge database of physiological and psychological changes, refine the index, and optimize the index data with people's social communicative symbols through the algorithm optimization. This is a huge and arduous long-term scientific research task and scientific research vision.

Bibliography

[1] ARIYATUM B, HOLLAND R, HARRISON D, et al. The future design direction of Smart Clothing development [J]. Journal of the Textile Institute, 2005, 96 (4): 199-210.

[2] BAKKER S, NIEMANTSVERDRIET K. The interaction-attention continuum: Considering various levels of human attention in interaction design [J]. International Journal of Design, 2016, 10: 1-14.

[3] BARNAGHI P, SHETH A, SINGH V, et al. Physical-cyber-social computing: Looking back, looking forward [J]. IEEE Internet Computing, 2015, 19 (3): 7-11.

[4] BARTHES R. Éléments de sémiologie [J]. Communications, 1964, 4 (1): 91-135.

[5] BRIGHOUSE H, SWIFT A. Family values: the ethics of parent-child relationships [M]. Princeton: Princeton University Press, 2014.

[6] BUCHANAN R. Human dignity and human rights: Thoughts on the principles of human-centered design [J]. Design Issues, 2001, 17 (3): 35-39.

[7] CHAN M, ESTÈVE D, FOURNIOLS J Y, et al. Smart wearable systems: Current status and future challenges [J]. Artificial Intelligence in Medicine, 2012, 56 (3): 137-156.

[8] CHATTARAMAN V, RUDD N A. Preferences for aesthetic attributes in clothing as a function of body image, body cathexis and body size [J]. Clothing and Textiles Research Journal, 2006, 24 (1): 46-61.

[9] CHEN M, MA Y J, LI Y, et al. Wearable 2.0: Enabling human-cloud integration in next generation healthcare systems [J]. IEEE Communications Magazine, 2017, 55 (1): 54-61.

[10] CHEN M, MA Y J, SONG J, et al. Smart clothing: Connecting human with clouds and big data for sustainable health monitoring [J]. Mobile Networks and Applications, 2016, 21 (5): 825-845.

[11] DE ACUTIS A, DE ROSSI D. E-garments: Future as "second skin"? [M] //Human-Computer Interaction Series. Cham: Springer International Publishing, 2017: 383-396.

[12] DUNLAP K. The development and function of clothing [J]. The Journal of General Psychology, 1928, 1 (1): 64-78.

[13] FEINBERG R A, MATARO L, BURROUGHS W J. Clothing and social identity [J]. Clothing and Textiles Research Journal, 1992, 11 (1): 18-23.

[14] GIACOMIN J. What is human centred design? [J]. The Design Journal, 2014, 17 (4): 606-623.

[15] GHAHREMANI HONARVAR M, LATIFI M. Overview of wearable electronics and smart textiles [J]. The Journal of the Textile Institute, 2017, 108 (4): 631-652.

[16] HARMS E. The psychology of clothes [J]. American Journal of Sociology, 1938, 44 (2): 239-250.

[17] KIM H Y, LEE J Y, MUN J M, et al. Consumer adoption of smart in-store technology: Assessing the predictive value of attitude versus beliefs in the technology acceptance model [J]. International Journal of Fashion Design, Technology and Education, 2017, 10 (1): 26-36.

[18] KIM Y K, WANG H, MAHMUD M S. Wearable body sensor network for health care applications [M] // Smart Textiles and their Applications. Amsterdam: Elsevier, 2016: 161-184.

[19] KING-O'RIAIN R C. Enduring or crossing distance for love? negotiating love and distance in the lives of mixed transnational couples [J]. Sociological Research Online, 2016, 21 (1): 151-160.

[20] KITTLER R, KAYSER M, STONEKING M. Molecular evolution of Pediculus humanus and the origin of clothing [J]. Current Biology: CB, 2003, 13 (16): 1414-1417.

[21] KOSTI M V, GEORGIADIS K, ADAMOS D A, et al. Towards an affordable brain computer interface for the assessment of programmers' mental workload [J]. International Journal of Human-Computer Studies, 2018, 115: 52-66.

[22] LAM PO TANG S, STYLIOS G K. An overview of smart technologies for clothing design and engineering [J]. International Journal of Clothing Science and Technology, 2006, 18 (2): 108-128.

[23] LEE J, BAGHERI B, KAO HUNG-AN. A Cyber-Physical Systems architecture for Industry 4.0-based manufacturing systems [J]. Manufacturing Letters, 2015, 3: 18-23.

[24] LIANG Y L, LEE S H, WORKMAN J E. Implementation of artificial intelligence in fashion: Are consumers ready? [J]. Clothing and Textiles Research Journal, 2020, 38 (1): 3-18.

[25] MALMIVAARA M. The emergence of wearable computing [M] //Smart Clothes and Wearable Technology. Amsterdam: Elsevier, 2009: 3-24.

[26] MARAZZITI D, BARONI S, GIANNACCINI G, et al. Decreased lymphocyte dopamine transporter in romantic lovers [J]. CNS Spectrums, 2017, 22 (3): 290-294.

[27] MASLOW A H. A theory of human motivation [J]. Psychological Review, 1943, 50 (4): 370-396.

[28] MORGADO M A. Animal trademark emblems on fashion apparel: A semiotic interpretation [J]. Clothing and Textiles Research Journal, 1993, 11 (3): 31-38.

[29] NAGAI Y. A sense of design: The embedded motives of nature, culture, and future [M] //TAURA T, ed. Principia Designae — Pre-Design, Design, and Post-Design. Tokyo: Springer Japan, 2014: 43-59.

[30] NAGAMACHI M. Kansei/affective engineering and history of kansei/affective engineering in the world [M] //Industrial Innovation. Boca Raton: CRC Press, 2010: 1-12.

[31] NING H S, LIU H, MA J H, et al. Cybermatics: Cyber–physical–social–thinking hyperspace based science and technology [J]. Future Generation Computer Systems, 2016, 56: 504-522.

[32] NONAKA I, TAKEUCHI H, UMEMOTO K. A theory of organizational knowledge creation [J]. International Journal of Technology Management, 2014, 11: 833-845.

[33] OWYONG Y S M. Clothing semiotics and the social construction of power relations [J]. Social

Semiotics, 2009, 19 (2): 191-211.

[34] PERRY A, MALININ L, SANDERS E, et al. Explore consumer needs and design purposes of smart clothing from designers' perspectives [J]. International Journal of Fashion Design, Technology and Education, 2017, 10 (3): 372-380.

[35] HJØRLAND B. Organizing knowledge. an introduction to managing access to information [J]. Journal of Documentation, 2009, 65 (1): 166-169.

[36] SHETH A, ANANTHARAM P, HENSON C. Physical-cyber-social computing: An early 21st century approach [J]. IEEE Intelligent Systems, 2013, 28 (1): 78-82.

[37] STEAD L, GOULEV P, EVANS C, et al. The emotional wardrobe [J]. Personal and Ubiquitous Computing, 2004, 8 (3): 282-290.

[38] STEEN M. Human-centered design as a fragile encounter [J]. Design Issues, 2012, 28 (1): 72-80.

[39] STOKES B, BLACK C. Application of the Functional, Expressive and Aesthetic Consumer Needs Model: Assessing the clothing needs of adolescent girls with disabilities [J]. International Journal of Fashion Design, Technology and Education, 2012, 5 (3): 179-186.

[40] TAY L, DIENER E. Needs and subjective well-being around the world [J]. Journal of Personality and Social Psychology, 2011, 101 (2): 354-365.

[41] TRINDADE I, MACHADO DA SILVA J, MIGUEL R, et al. Design and evaluation of novel textile wearable systems for the surveillance of vital signals [J]. Sensors, 2016, 16 (10): 1573.

[42] VAN LANGENHOVE L, HERTLEER C. Smart clothing: A new life [J]. International Journal of Clothing Science and Technology, 2004, 16 (1/2): 63-72.

[43] VEHMAS K, RAUDASKOSKI A, HEIKKILÄ P, et al. Consumer attitudes and communication in circular fashion [J]. Journal of Fashion Marketing and Management, 2018, 22 (3): 286-300.

[44] WRIGHT R, KEITH L. Wearable technology: If the tech fits, wear it [J]. Journal of Electronic Resources in Medical Libraries, 2014, 11 (4): 204-216.

[45] WANG W Z, FANG Y, Nagai Y, et al. Integrating Interactive Clothing and Cyber-Physical Systems: A Humanistic Design Perspective [J]. Sensors, 2019, 20 (1):127.

[46] WANG W Z, WANG S Y. Toward parent-child smart clothing: Purchase intention and design elements [J]. Journal of Engineered Fibers and Fabrics, 2021, 16 (4):259-273.

[47] WANG W Z, ZOU J W, FANG Y. Design and Evaluation of a Somatosensory Hat: An Emotional Semantic Perspective [J]. AATCC Journal of Research, 2021, 8:20-29.

[48] WANG W Z, NAGAI Y, FANG Y, et al. Interactive technology embedded in fashion emotional design [J]. International Journal of Clothing Science and Technology, 2018, 30 (3): 302-319.

[49] WANG W Z, WANG Y, YU S L, et al. Design for mutual transformation between outdoor wear and camping tent [J]. International Journal of Clothing Science and Technology, 2014, 26 (4): 291-304.